RESEARCH NEEDS FOR HIGH-LEVEL WASTE STORED IN TANKS AND BINS AT U.S. DEPARTMENT OF ENERGY SITES

ENVIRONMENTAL MANAGEMENT SCIENCE PROGRAM

Committee on Long-Term Research Needs for
Radioactive High-Level Waste at
Department of Energy Sites
Board on Radioactive Waste Management
Division on Earth and Life Studies

National Research Council

NATIONAL ACADEMY PRESS
Washington, D.C.

NOTICE: The project that is the subject of this report was approved by the Governing Board of the National Research Council, whose members are drawn from the councils of the National Academy of Sciences, the National Academy of Engineering, and the Institute of Medicine. The members of the committee responsible for the report were chosen for their special competence and with regard for appropriate balance.

Support for this study was provided by the U.S. Department of Energy, under Grant No. DE-FC01-99EW59049. All opinions, findings, conclusions, and recommendations expressed herein are those of the authors and do not necessarily reflect the views of the Department of Energy.

International Standard Book Number: 0-309-07565-3

Additional copies of this report are available from:

National Academy Press
2101 Constitution Avenue, N.W.
Box 285
Washington, DC 20055
800-624-6242
202-334-3313 (in the Washington Metropolitan Area)
http://www.nap.edu

Cover: The photographs in the background show one bin containing calcined high-level waste at the Idaho Engineering and Environmental Laboratory (upper left) and six tanks at the Hanford Site before they were buried (lower right).

Printed in the United States of America.

THE NATIONAL ACADEMIES

National Academy of Sciences
National Academy of Engineering
Institute of Medicine
National Research Council

The **National Academy of Sciences** is a private, nonprofit, self-perpetuating society of distinguished scholars engaged in scientific and engineering research, dedicated to the furtherance of science and technology and to their use for the general welfare. Upon the authority of the charter granted to it by the Congress in 1863, the Academy has a mandate that requires it to advise the federal government on scientific and technical matters. Dr. Bruce Alberts is president of the National Academy of Sciences.

The **National Academy of Engineering** was established in 1964, under the charter of the National Academy of Sciences, as a parallel organization of outstanding engineers. It is autonomous in its administration and in the selection of its members, sharing with the National Academy of Sciences the responsibility for advising the federal government. The National Academy of Engineering also sponsors engineering programs aimed at meeting national needs, encourages education and research, and recognizes the superior achievements of engineers. Dr. Wm. A. Wulf is president of the National Academy of Engineering.

The **Institute of Medicine** was established in 1970 by the National Academy of Sciences to secure the services of eminent members of appropriate professions in the examination of policy matters pertaining to the health of the public. The Institute acts under the responsibility given to the National Academy of Sciences by its congressional charter to be an adviser to the federal government and, upon its own initiative, to identify issues of medical care, research, and education. Dr. Kenneth Shine is president of the Institute of Medicine.

The **National Research Council** was organized by the National Academy of Sciences in 1916 to associate the broad community of science and technology with the Academy's purposes of furthering knowledge and of advising the federal government. Functioning in accordance with general policies determined by the Academy, the Council has become the principal operating agency of both the National Academy of Sciences and the National Academy of Engineering in providing services to the government, the public, and the scientific and engineering communities. The Council is administered jointly by both Academies and the Institute of Medicine. Dr. Bruce Alberts and Dr. Wm. A. Wulf are chairman and vice-chairman, respectively, of the National Research Council.

COMMITTEE ON LONG-TERM RESEARCH NEEDS FOR HIGH-LEVEL WASTE AT THE DEPARTMENT OF ENERGY SITES

MICHAEL CORRADINI, Chair, University of Wisconsin, Madison
DAVID CAMPBELL, Oak Ridge National Laboratory (retired), Oak Ridge, Tennessee
MICHELINE DRAYE, École Nationale Supérieure de Chimie de Paris, France
CHARLES DRUMMOND III, Ohio State University, Columbus
PETER HAYWARD, Eutechnics Consulting, Inc., Pinawa, Manitoba, Canada
LINN HOBBS, Massachusetts Institute of Technology, Cambridge
EDWARD LAHODA, Westinghouse Electric Science and Technology Department, Pittsburgh, Pennsylvania
ROBIN ROGERS, The University of Alabama, Tuscaloosa
BEN STERNBERG, University of Arizona, Tucson
EDWIN ZEBROSKI, Elgis Consulting, Inc., Los Altos, California

LIAISON TO THE BOARD ON RADIOACTIVE WASTE MANAGEMENT

ROBERT M. BERNERO, U.S. Nuclear Regulatory Commission (retired), Gaithersburg, Maryland

STAFF

BARBARA PASTINA, Staff Officer
LAURA D. LLANOS, Senior Project Assistant
MATTHEW BAXTER-PARROTT, Project Assistant

List of Report Reviewers

This report has been reviewed in draft form by individuals chosen for their diverse perspectives and technical expertise, in accordance with procedures approved by the National Research Council Report Review Committee. The purpose of this independent review is to provide candid and critical comments that will assist the institution in making its published report as sound as possible and to ensure that the report meets institutional standards for objectivity, evidence, and responsiveness to the study charge. The review comments and draft manuscript remain confidential to protect the integrity of the deliberative process. We wish to thank the following individuals for their review of this report:

Philip Clark, Sr., GPU Nuclear Corporation (retired)
William J. Hinze, Purdue University
Milton Levenson, Bechtel International (retired)
Werner Lutze, University of New Mexico
Dale Niesz, Rutgers University
Martin Steindler, Argonne National Laboratory
Victoria J. Tschinkel, Landers and Parsons
Raymond G. Wymer, Oak Ridge National Laboratory (retired)

Although the reviewers listed above have provided many constructive comments and suggestions, they were not asked to endorse the conclusions or recommendations nor did they see the final draft of the report before its release. The review of this report was overseen by Clarence Allen of the California Institute of Technology. Appointed by the National Research Council, he was responsible for making certain that an independent examination of this report was carried out in accordance with institutional procedures and that all review comments were carefully considered. Responsibility for the final content of this report rests entirely with the authoring committee and the institution.

Preface

The objective of this study is to provide recommendations to the U.S. Department of Energy's (DOE's) Environmental Management Science Program on the development of a long-term basic research agenda to address problems in the management of high-level waste (HLW) stored in tanks or bins at DOE sites. This report does not address other HLW-related problems.

The complete statement of task is shown in Sidebar P.1 below. This study was sponsored by DOE's Office of Environmental Management. To accomplish this project, the National Research Council empanelled a 10-member committee within the Board on Radioactive Waste Management. The committee members were chosen for their expertise in relevant domains such as HLW management issues at DOE sites, risk assessment, nuclear and chemical engineering, radiation and analytical chemistry, materials science and engineering, HLW processing and

SIDEBAR P.1 STATEMENT OF TASK

The objective of this study is to provide recommendations to the Department of Energy's Environmental Management (EM) Science Program on the development of a long-term basic research agenda to address HLW problems at DOE sites. The report will accomplish the following:

- **Identify significant HLW problems that cannot be addressed effectively with current technologies.**
- **Recommend areas of research where the EM Science Program can make significant contributions to solving these problems and adding to scientific knowledge generally.**

In recommending specific areas of research, the committee should take into account, where possible, the agendas of other HLW-related research programs. The committee may also consider and make recommendations, as appropriate, on the processes by which (1) future research needs can be identified and (2) successful research results can be applied to DOE's HLW problems.

immobilization forms, and geotechnical engineering. Two of the committee members were chosen from foreign countries (France and Canada) with HLW and/or spent nuclear fuel management programs.

This study could not have been completed without the help of various individuals and organizations. The committee would especially like to thank Carolyn Huntoon, Gerald Boyd, Mark Gilbertson, and Ker-Chi Chang of DOE's Office of Environmental Management for their continued support and assistance throughout this study. The committee acknowledges the hard work of the individuals who organized site tours and briefings. In particular, the committee thanks Don Woodrich, (Handford Site), Jerome Morin, Thomas Gutmann, and Loucien Papouchado (Savannah River Site). The committee also recognizes the efforts of all the speakers who gave presentations during the information-gathering phase of this study (see Appendix F).

The following individuals provided additional information to the committee: Carol Jantzen, Ned Bibler, and Walter Tamosaitis (Savannah River Site); Kenneth Picha, Theodore Pietrok, Kurt Gerdes, Joe Cruz, and Pete Gibbons (Tanks Focus Area); Greg Mitchem, and Tony Knepp (Hanford Site); Ian Pegg and Isabelle Muller (The Catholic University of America); Robert Perdue (Westinghouse Science and Technology Department); Yves Poitevin (Commissariat à l'Energie Atomique, France); Jean-Paul Moulin (Socitété Générale pour les Techniques Nouvelles, France); and Maurice Tarnero (Compagnie Générale des Matières Nucléaires, France).

The committee also thanks the following staff of the National Research Council's Board on Radioactive Waste Management for their support, dedication, and hard work in guiding the report from its early stages through publication: Barbara Pastina, Kevin Crowley, Laura Llanos, and Matthew Baxter-Parrot.

This report reflects a consensus of the committee and has been reviewed in accordance with the procedures of the National Research Council. As is the normal practice of the National Academies, committee members do not represent the views of their institutions, but form an independent body to author this report.

Michael Corradini
Chairman
May 2001

Contents

Executive Summary

The United States Department of Energy (DOE) has approximately 400 million liters (100 million gallons) of liquid high-level waste (HLW) stored in underground tanks and approximately 4,000 cubic meters of solid HLW stored in bins. The current DOE estimate of the cost of converting these liquid and solid wastes into stable forms for shipment to a geological repository exceeds $50 billion to be spent over several decades (DOE, 2000).

This committee[1] was appointed by the National Research Council (NRC) to advise the Environmental Management Science Program (EMSP)[2] on a long-term research agenda addressing the above problems related to HLW[3] stored in tanks and bins at DOE sites.[4] The complete statement of task is reproduced in Sidebar P.1 of the Preface.

The committee outlined HLW problem areas that either cannot be addressed effectively with current technologies, or that are anticipated for the future. From these problem areas, the committee identified research objectives, leaving to EMSP investigators the role of determining the pathway to achieving these objectives through basic research. Moreover, it is not the purpose of this report to circumscribe the investigators' creativity by giving a detailed list of research projects.

---

[1]Committee on Long-Term Research Needs for Radioactive High-Level Waste at Department of Energy Sites.

[2]The EMSP currently provides the DOE's Office of Environmental Management with a basic research program to reduce costs, time, and risks associated with cleaning up the nation's nuclear complex.

[3]HLW is the highly radioactive waste from the chemical reprocessing of spent fuel and target materials to recover plutonium and uranium for the production of nuclear weapons. The committee did not address issues related to spent nuclear fuel, transuranic waste, in particular plutonium, or other secondary waste streams from the processing and handling of HLW (see Chapter 1).

[4]The HLW sites are the Hanford Site, Washington, the Savannah River Site, South Carolina, the Idaho Engineering and Environmental Laboratory, and the West Valley Demonstration Project, New York.

Throughout the report, examples of research topics are provided to illustrate how the committee would address the identified HLW problems. These examples are not to be considered as recommendations or research priorities.

Long-Term Research Needs for HLW Management

The recommended long-term research agenda is organized according to DOE's current approach to HLW management as follows:

- Characterization
- Retrieval and pretreatment
- Immobilization
- Tank closure

The motivation for selecting the long-term basic research activities in this report is to provide contingency approaches for DOE's HLW management programs and to improve process effectiveness. The committee recommends research topics that would *provide contingency approaches* as the basis for program support in case of interferences or disruptions to current HLW management plans. Results from these research activities would help reduce technological risk. The committee also recommends research topics that would *improve process effectiveness.*

The long-term basic research activities are summarized below in terms of research objectives. More details can be found in the chapters indicated in parenthesis.

Characterization (Chapter 3): **Innovative methods to achieve real-time and, when practical, in situ characterization data for HLW and process streams that could be useful for all phases of the waste management program.** The objective of research activities in the area of characterization is to provide the scientific basis for the following:

- developing remote sensing instruments, and
- developing on-line or in situ instruments.

Pretreatment (Chapter 4): **High-efficiency, high-throughput separation methods that could reduce HLW program costs over the next sev-**

eral decades. The objective of research activities in the area of pretreatment is to provide the scientific basis for the following:

- improving solids-liquid separation methods,
- improving underlying science and technology for sludge leaching operations, and
- increasing the efficiency of liquid decontamination methods.

Immobilization (Chapter 5): **Robust, high-loading immobilization methods and materials that could provide enhancements or alternatives to the current immobilization strategy.** The objective of research activities in the area of immobilization is to provide the scientific basis for the following:

- identifying alternative immobilization media to overcome limitations of borosilicate glasses,
- investigating the effect of increased crystal content on the durability of the borosilicate glass matrix,
- improving phenomenological models to predict long-term leachability of various glass waste forms,
- evaluating advantages of using unreacted glass-forming chemicals compared to premelted glass frit,
- understanding foaming in Joule-heated melters,
- mitigating the effects of precipitation of noble metals and crystalline phases in Joule-heated melters,
- improving Joule-heated melters to achieve higher processing temperatures, and
- developing alternatives to Joule-heated melting.

Tank closure and other long-term issues (Chapter 6): **Innovative methods to achieve tank closure, non-invasive monitoring of the near-field areas, as well as improved containment barriers.** The objective of research activities on this topic is to provide the scientific basis for the following:

- developing highly innovative and effective retrieval methods for removal of residual HLW materials from tanks,
- developing highly innovative and effective retrieval methods for removal of HLW materials from connecting pipelines,
- improving the characterization of residual waste in tanks,
- developing noninvasive near-field monitoring techniques, and
- improving near-field containment methods through the use of barriers.

EMSP Research Portfolio: Technological Risk and Desired Attributes

The committee provides programmatic recommendations on the management of technological risk and discusses desired attributes for the EMSP research portfolio. An analysis of this portfolio, with respect to HLW research activities, is provided as an interim report, reproduced in Appendix A. The main findings and recommendations from Appendix A are included in the body of this report.

Managing Technological Risk

The research program proposed by the committee could help the EMSP in reducing technological risk.[5] Technological risk can be reduced by improving process effectiveness and by providing contingency approaches to baseline technologies before disruptions or interferences to current HLW processing programs occur. Improving process effectiveness can lead to more robust approaches that require fewer steps to achieve the desired result; therefore, a simple and robust process presents less technological risk compared to a multi-step process. This is because there is a finite probability of encountering a problem causing process failure in any one of the individual steps. For instance, separation techniques removing more than one targeted species in a single step pose fewer technological risks than a series of steps to achieve the same result. To provide contingency approaches, it is necessary to allocate funding for exploratory research relevant to HLW cleanup. The committee recognizes that exploratory research efforts do not bear fruit immediately and that results often are not deployed in the field. However, the success of a basic research program should not be measured only by the proportion of projects that become field-applicable.

Desired Attributes of the EMSP Research Portfolio

In selecting the projects for future HLW proposal cycles, attention to the following programmatic recommendations is also warranted:

1. *Maintain long-term vision.* Research projects should be focused not only on DOE's short-term issues but also on significant long-

---

[5]A technological risk in HLW management is the risk that existing technologies will fail to accomplish goals and performance requirements set by environmental remediation policies or regulations.

term problems to advance the state of knowledge well beyond the next decade. This approach maintains the EMSP long-term mission and mitigates the technological risk that would interrupt the Office of Environmental Management (EM) HLW program.

2. *Maintain relevance.* The EMSP should support basic research on processes and phenomena that support the EM program mission. These projects can then feed into other EM programs and spark the applied research and development that will be needed for implementation at the sites. Synergies between the EMSP and other research programs are natural and should be exploited to their fullest extent. Collaborations with foreign countries with relevant HLW research activities should be encouraged. At a minimum, the EMSP should be aware of scientific results and research trends for HLW management in countries with similar problems.
3. *Provide for contingencies.* The EMSP should promote underlying science and technology that will support contingency approaches to address unanticipated difficulties encountered in baseline processes. Some fraction of EMSP projects should support exploratory and innovative research involving non-conventional technologies, possibly leveraged from other disciplines. The projects should represent a balanced range of research styles, from large-scale teams to single investigators.
4. *Develop working relationships.* EMSP investigators should interact with problem holders at the sites to learn about the nature of the problems to be solved. In return, problem holders might gain a better understanding of the scientific gaps underlying HLW problems. EMSP researchers could visit DOE sites regularly to learn more about specific issues. Conversely, EMSP could identify "liaisons" among the problem holders at the sites to communicate with its investigators. The liaisons will not only have the greatest knowledge about the sites but will also be able to assist in integrating the results of EMSP research into the long-term EM effort.
5. *Prioritize objectives.* The EMSP should prioritize the two research *objectives* of providing contingency approaches and improving process effectiveness, as follows. The EMSP should strive for a *balanced portfolio* addressing both *already identified problems* and *future potential problems.* The portfolio should have *a primary focus on identified problems* that must be solved either for efficient HLW processing operations or to provide for contingencies. Characterization, separation, immobilization methods and processes are problem areas where EMSP research could improve operations and provide immediate support in case of

interferences or disruptions to current HLW processing programs. At the same time, there should also be a *consideration of potential problems* that could arise or become exacerbated in the future. Examples of recommended research to address future potential problems are the following: new or better separation techniques to remove bulk non-hazardous material from the HLW stream, methods to achieve higher waste loading in immobilized waste forms, and improvements in tank closure and near-field monitoring issues.

The committee recognizes that the EMSP cannot address all of the recommended research areas equally in the next few years, nor can its portfolio acquire all the recommended program attributes immediately. The committee is also aware of DOE's commitments and of the difficulties in implementing changes, both from a technology and regulatory point of view. However, current plans to treat and dispose of HLW are fraught with technical uncertainties, and many of the planned treatment activities are first-of-a-kind efforts presenting enormous technical challenges. Given the long-term duration of this planned clean-up effort, it is expected that new technologies will emerge and that greater scientific understanding will be achieved in the next decades. Therefore, the EMSP is the ideal setting to develop truly innovative approaches for the management of HLW that could lead to scientific breakthroughs in the future. If a scientific breakthrough can help reduce risks, cleanup time, and costs, then regulations can be revised accordingly and obstacles to implementing changes at the sites can be removed. In this respect, the EMSP is a "small but vital element" to the long-term success of the EM cleanup mission.

1

Introduction

"The Department [of Energy] has about 100 million gallons of liquid HLW [high-level waste] stored in underground tanks at its Savannah River and Hanford Sites and about 4,000 cubic meters of solid HLW stored in bins at the Idaho Site. The Department estimates that it will spend on the order of $50 billion over the next 50 years or so to put this waste into a more stable form for shipment to a geological repository. [...] The Department believes that it must invest in long-term basic research to address effectively its HLW [...] responsibilities and would like to have the National Research Council's advice on what research investments should be made" (Huntoon, 1999).

In response to this request, the present study was undertaken to assist the U.S. Department of Energy's (DOE's) Office of Environmental Management (EM) in the development of a long-term basic research agenda (see Sidebar 1.1) addressing the HLW problems at DOE sites set out above. HLW is defined by DOE as the highly radioactive waste from the chemical reprocessing of spent fuel and target materials to recover plutonium and uranium for the production of nuclear weapons (Sidebar 1.2). The EM Science Program (EMSP) provides the basic research support for environmental cleanup activities across DOE's HLW sites. The EMSP mission and other details on the program are given in Sidebar 1.3 and in Appendix D.

Objectives of This Study

As mentioned in the statement of task (Sidebar P.1 of the Preface), the objective of this study is twofold:

1. identify significant HLW problems that cannot be addressed effectively with current technologies, and

SIDEBAR 1.1 COMMITTEE'S DEFINITION OF LONG-TERM BASIC RESEARCH FOR THE EMSP

For the purpose of this study, the committee interpreted the expression *"long-term basic research,"* mentioned in the statement of task, as research that creates new generic knowledge underlying fundamental processes and phenomena and is focused on long-term, rather than short-term, problems. Moreover, in alignment with the EMSP mission, this research should be *"needs-driven"* or *"mission-directed,"* with potential high-impact results on the EM-HLW management program. A similar definition of basic research was provided in other NRC reports relevant to the EMSP (NRC, 1997, 2000a, 2000c). This definition differs from that used by academia and by government agencies, such as the National Science Foundation, where basic research is "research whose objective is to gain knowledge or understanding of phenomena without specific applications in mind" (AAAS, 2001).

2. recommend areas of research where the EMSP can make significant contributions to solve these problems and increase scientific knowledge generally.

The ultimate objective of the committee's recommendations is to help EMSP build a long-term science program that will contribute to the EM cleanup mission of reducing risks to the environment, workers and public, and reducing cleanup time and costs.

Strategy To Address the Task

The recommended long-term research agenda is organized according to DOE's current approach for HLW management[1] as follows:

- Characterization
- Retrieval and pretreatment
- Immobilization
- Tank closure

The motivation for selecting the long-term research activities in this report is to *provide contingency approaches* for DOE's HLW management programs and to *improve process effectiveness.* The committee recommends research topics that would provide contingency approaches as the basis for program support in case of interferences or disrup-

[1]This approach is discussed in more detail in Chapter 2.

tions to current waste management plans. Results from these research activities would contribute to reducing technological risk over the next decades. The committee also recommends research topics that would improve process effectiveness or lead to a decrease in the volume of immobilized HLW and secondary waste streams.

The long-term research activities recommended in this report address both *identified* problems that are already affecting the HLW sites cleanup and *future potential problems*. In addressing the task of developing a long-term basic research agenda, the committee identified *problem areas* and *research objectives*, rather than specific research projects. It is not possible, with today's knowledge, to clearly define specific research activities addressing many of DOE's HLW problems. It is the role of EMSP investigators to determine the pathway to achieve these research objectives as more information becomes available. The results of the EMSP research could potentially find application at any time during the multi-decade EM cleanup program.

Because the EMSP research program should be needs-driven (see Sidebar 1.1), the committee identified broad problem areas throughout the HLW management process. It is not the purpose of this report to circumscribe investigators' creativity by giving a detailed list of research projects; rather, the committee leaves to the investigators the charge of addressing the scientific basis underlying the problem areas described

SIDEBAR 1.2 DEFINITION OF HIGH-LEVEL WASTE

High-level waste was defined in the Nuclear Waste Policy Act (1982) as

(A) the highly radioactive material resulting from the reprocessing of spent nuclear fuel, including liquid waste produced directly in reprocessing and any solid material derived from such liquid waste that contains fission products in sufficient concentrations; and

(B) other highly radioactive material that the U. S. Nuclear Regulatory Commission (USNRC), consistent with existing law, determines by rule requires permanent isolation.

The definition from the 1982 Nuclear Waste Policy Act has been amplified (Congress, 1982). The USNRC has defined, in Title 10 Code of Federal Regulations (CFR) Part 60.2, that HLW "means: (1) Irradiated reactor fuel, (2) liquid wastes resulting from the operation of the first cycle solvent extraction system, or equivalent, and the concentrated wastes from subsequent extraction cycles, or equivalent, in a facility for reprocessing irradiated reactor fuel, and (3) solids into which such liquid wastes have been converted." Commercial spent fuel is stored for direct disposal and is classified as HLW according to the USNRC definition. However, according to DOE's definition, irradiated reactor fuel is not classified as HLW but as "spent fuel," pending a decision to dispose of it, as explained in DOE's Order 435.1 on radioactive waste management, and its implementation guide G.435.1-1 (DOE, 1999, pages II-1 to II-6).

SIDEBAR 1.3 THE EMSP AND ITS MISSION

The EMSP, created in 1996, is a partnership between DOE's EM and SC. The mission of the EMSP is the following (DOE-EMSP, 2000):

- **Develop a targeted, long-term basic research agenda to reduce cleanup costs and risks to workers and public;**
- **Bridge the gap between broad fundamental research and needs-driven applied technology; and**
- **Serve as a stimulus for focusing the nation's science infrastructure on critical environmental problems.**

The EMSP directs research in seven EM "problem areas": HLW, deactivation and decommissioning, spent nuclear fuel, nuclear materials, subsurface contamination, mixed and transuranic waste, and health/ecology/risk. The EMSP develops its research program through complex-wide surveys, site workshops, system analysis, and recently, through the advice of the National Research Council (NRC, 2000a, 2000c). EMSP grants are awarded to research projects responding to requests for proposals (RFPs), published in the Federal Register, issued by the EM and the SC. The proposals are screened through a two-step peer review process, described in Appendix D.

in this report. Therefore, the recommendations are provided under the form of research objectives warranting further basic research investment, rather than research activities. Throughout the report, examples of research topics are provided to illustrate how the committee would address the identified HLW problems. These examples are not to be considered as recommendations or research priorities.

Some of the issues discussed in this report were not considered as "high priority" for basic research as others because different programs within DOE are already addressing them. Moreover, the committee did not want to "drown" high-priority research in an excessively long list of needs.

Boundaries of the Statement of Task

The committee has elected to focus its attention and information-gathering activities on HLW problems involving only wastes stored in tanks and bins at the principal DOE HLW sites.[2] Consistent with the statement of task, this study does not address issues related to spent

[2]This decision has been made in agreement with the sponsor during the first committee meeting.

nuclear fuel, transuranic waste (in particular plutonium), or other secondary waste streams from the processing and handling of HLW. Appendix C gives a brief perspective on DOE's spent nuclear fuel issues. There are several other problems related to HLW at DOE sites. These problems have been or are being discussed by other NRC (National Research Council) committees. For instance, three NRC committees accomplished or are currently accomplishing the task of identifying research needs for the EMSP in the subsurface contamination (NRC, 2000a), deactivation and decommissioning (NRC, 2000c, 2001b), and transuranic and mixed waste[3] problem areas. Non-radioactive wastes (in particular, toxic inorganic materials and dense non-aqueous phase liquids) are addressed in two previous NRC reports: Research Needs in Subsurface Science (NRC, 2000a), and Groundwater and Soil Cleanup: Improving Management of Persistent Contaminants (NRC, 1999d).

Long-term institutional management issues, such as site monitoring after tank closure, have been briefly addressed in this report. Long-term research needs and other broader issues related to institutional management, such as the problem of data storage, are addressed in a more comprehensive two-phase NRC study entitled Long-Term Institutional Management of U.S. DOE Legacy Waste Sites (NRC, 2000d). Finally, there are many questions related to the proposed HLW repository. These issues are the territory of DOE's Office of Civilian Radioactive Waste Management (OCRWM), not the EM; therefore, they are beyond this statement of task and are not addressed in this report.

Committee's Methodology

During this study, the committee met six times to gather information, review relevant literature, and develop this report. DOE and its contractors fully cooperated with the committee during the three information-gathering meetings and follow-up questions on the four major HLW sites in the country (the Hanford Site, the Savannah River Site, the Idaho Engineering and Environmental Laboratory, and the West Valley Demonstration Project). The committee also visited the Hanford Site and the Savannah River Site. The list of briefings received during the open-session meetings is reported in Appendix F.

Relevant literature reviewed by the committee included previous NRC reports focusing on the EMSP (NRC, 1997a, 2000a) and on HLW management at DOE sites (NRC, 1999a, 1999b, 2000b). In recommending research topics, the committee considered the EMSP basic

---

[3]This report is expected in Fall 2002.

research program plan (DOE-EMSP, 2001) in relation to the agendas of more applied research and development activities encompassed by other HLW-related research programs within the EM. The committee also considered research activities in HLW management taking place in foreign countries.

At the halfway point of this study, the committee released an interim report, reproduced in Appendix A, to address selected issues meriting early attention and action by the EMSP during the preparation of a request for proposals (RFP) for HLW-related research. The main findings and recommendations from Appendix A are included in the body of this report. DOE's Office of Science (SC), in collaboration with EM, published the RFP in January 2001 (Federal Register, 2001) and both offices were reviewing the proposals received when this report was released. Since the interim report, the committee gathered additional information and approached HLW cleanup issues from a broader perspective to fully address the statement of task.

Organization of This Report

Chapter 2 presents an overview of HLW management challenges at DOE sites. Chapters 3 through 6 identify issues and research needs in the HLW management process areas (waste characterization, retrieval, pretreatment,[4] immobilization, tank closure, and other long-term issues). Chapter 7 addresses the issue of technological risk, and provides some programmatic considerations for the EMSP research portfolio. Chapter 8 presents a summary (in Table 8.1) of the long-term basic research needs, listed according to their objectives and respective process areas. Table 8.1 indicates whether the purpose of the research activity is to provide contingency approaches or to improve process effectiveness. The table also indicates whether each research activity addresses "already identified" or "future potential" problems.

---

[4]Waste pretreatment (see Chapter 4) consists of a series of processes to separate constituents that are radioactive, hazardous, or detrimental for the immobilization step from the bulk of the waste.

2

Overview of HLW Challenges at DOE Sites

The DOE has approximately 400 million liters (100 million gallons) of liquid HLW stored in underground tanks and approximately 4,000 cubic meters of solid HLW stored in bins. The current DOE estimate of the cost of converting these liquid and solid wastes into stable forms for shipment to a geological repository exceeds $50 billion to be spent over several decades (DOE, 2000).

The most important HLW sites in the United States are the following:

- the Hanford Site, near Richland, Washington,
- the Idaho National Engineering and Environmental Laboratory (INEEL), near Idaho Falls,
- the Savannah River Site (SRS), near Aiken, South Carolina, and
- the West Valley Demonstration Project (WVDP), near Buffalo, New York.

All of the HLW sites listed above are briefly described later in this chapter.

Sidebar 2.1 shows a simplified diagram of one of the processes that produced HLW during chemical recovery of plutonium and uranium from spent fuel and target materials for the production of nuclear weapons. Most of the HLW is a multiphase mixture of solids and liquids stored in underground tanks. Some has been retrieved and solidified and is stored in underground bins or surface vaults. Processing and handling of these HLW forms generate other radioactive wastes managed as transuranic or low-level waste (LLW, see Sidebar 2.2), depending on their characteristics. The reclassification of HLW residues to LLW or transuranic takes place according to DOE's procedures to determine

SIDEBAR 2.1 THE PLUTONIUM AND URANIUM EXTRACTION (PUREX) PROCESS

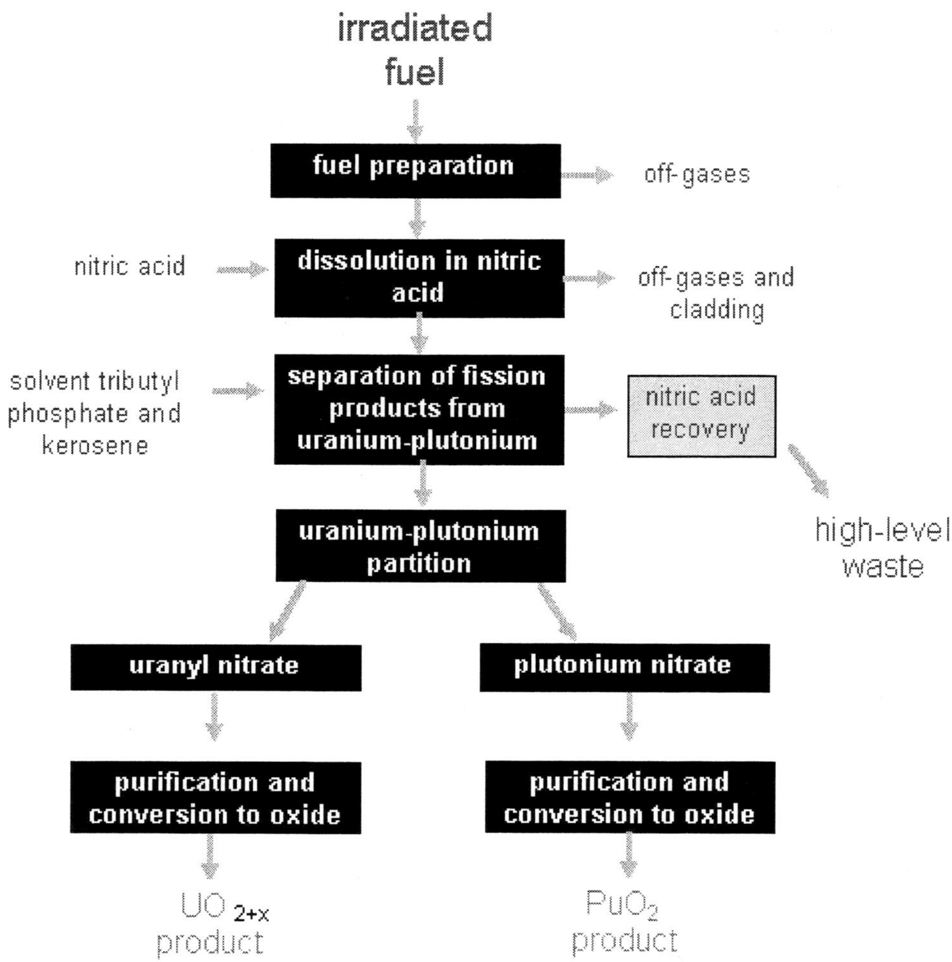

The PUREX process, shown above, purified and separated both uranium and plutonium from defense spent nuclear fuel. It was first used at the Hanford Site in 1952 and subsequently at the other HLW sites. The first step consists of removing the cladding of spent fuel (fuel preparation) to expose fuel rods. In the second step, spent fuel is dissolved in nitric acid. The third and key step is the separation and recovery of uranium and plutonium from the fission products. It is done in a continuous counter-current solvent extraction process using tri-N-butyl phosphate in a hydrocarbon diluent (such as kerosene). Plutonium and uranium are extracted in the organic solvent while fission products and other impurities remain dissolved in the highly acidic aqueous nitric solution stored in tanks and classified as HLW. The fourth step consists of separating plutonium from uranium by reducing plutonium to the organic-insoluble, trivalent state while uranium remains in the organic phase in the oxidized form. Both plutonium and uranium are then converted to solid oxide or metal. Except at the INEEL, the

waste was neutralized with sodium hydroxide prior to storage in the tank farms. A major advantage of the PUREX process over the earlier processes used (bismuth phosphate and REDOX processes, see the glossary in Appendix G) is that nitric acid is the only salting agent used in the aqueous phase; therefore HLW has a lower salt content than waste from other processes. Other advantages of PUREX are reduction of the waste volume, greater flexibility in process conditions, reduction of fire and other hazards, and reduction of costs. Processes based on PUREX have been adopted for nearly all fuel reprocessing throughout the world. SOURCE: Adapted from Benedict et al. (1981) and from IEER (1996).

whether the waste can be treated as waste incidental to reprocessing (WIR) described in Chapter 6, Sidebar 6.1. The committee has not considered issues with these other wastes in order to focus its attention on the retrieval and processing of HLW from tanks and bins.

HLW Management Strategy at DOE Sites

The current DOE plans to treat and dispose of HLW face many technical uncertainties. Many of the planned treatment activities are first-of-a-kind efforts involving highly radioactive liquid and solid wastes. Moreover, the waste streams are fundamentally different in character, both among the HLW sites and within the sites, because waste streams were generated from processing different kinds of nuclear fuel and were treated and stored differently. Consequently, each HLW stream presents different challenges and may require specific processing and immobilization strategies.

However, the overall HLW management strategy applied by DOE is the same throughout all HLW sites. The generic HLW remediation flow sheet is shown in Figure 2.1 and comprises the "backbone" of this report. This flow sheet applies mainly to HLW in the tanks. In the case of HLW stored at the INEEL in a calcined form, some details are slightly different, as described later in this chapter. The HLW management strategy consists of different steps, described as follows.

Characterization of waste in the tanks is performed to ensure the continued safe storage of the waste and to facilitate the following HLW management operations. At a given point, HLW must be retrieved from the storage tanks and bins and processed for ultimate disposal in a repository. The process of removing waste from the tanks and transporting it to the processing facilities is called *retrieval*. After retrieval, the HLW proceeds to the *pretreatment* phase, to separate constituents that

SIDEBAR 2.2 NUCLEAR AND RADIOACTIVE WASTE TERMS

A. High-level waste: The radioactive by-product generated from processing irradiated fuel to separate usable plutonium and other isotopes for weapons, research, and new fuel; and irradiated fuel itself if destined for direct disposal.

B. Transuranic waste: Waste that contains alpha-emitting transuranic elements with half-lives greater than 20 years whose combined activity level is at least 100 nanocuries (3,700 becquerels) per gram of waste at the time of assay.

C. Low-level waste: All radioactive waste not classified as HLW, transuranic waste, spent nuclear fuel, or the by-product material (see definition F.3.b below) generated by the processing of ores for extraction of natural uranium or thorium.

D. Mixed waste: Waste that contains both radioactive material regulated under the Atomic Energy Act and hazardous chemical compounds regulated under the Resource Conservation and Recovery Act, such as mercury, polychlorinated biphenyls, or organic solvents.

E. Spent nuclear fuel: Nuclear fuel that has been withdrawn from a nuclear reactor following irradiation, has undergone at least one year's decay since being used as a source of energy in a power reactor, and has not been chemically separated into its constituent elements by reprocessing. Spent nuclear fuel, through fission and neutron activation and decay, contains multiple radioactive elements with varying chemical and radiological properties. Spent fuel includes the special nuclear material, byproduct material, source material, and other radioactive materials associated with fuel assemblies.

F. Nuclear material (Atomic Energy Act material): Nuclear material generally includes (1) *source material,* natural uranium, thorium, or any other material that, alone or in combination, is determined to be a potential fuel for a fission reactor; (2) *special nuclear material,* fissile material such as uranium enriched in uranium-235, or plutonium that can undergo fission; and (3) *byproduct material,* meaning (a) any radioactive material (except *special nuclear material*) yielded in or made radioactive by exposure to the radiation incident to the process of producing or utilizing *special nuclear material,* and (b) the tailings or wastes produced by the extraction or concentration of uranium or thorium from any material processed primarily for its *source material* content.

G. Radioactive material: Radioactive material includes all of the nuclear material defined above. Radioactive material also includes naturally occurring radioactive elements such as isotopes of radium as well as material made radioactive by processes other than nuclear fission.

SOURCE: Adapted from DOE (2000).

are radioactive, hazardous, or detrimental to the immobilization step from the bulk of the waste. After pretreatment, HLW is ready for *immobilization* in a borosilicate glass matrix. This process is called "vitrification." The HLW glass canisters produced are stored at the sites until

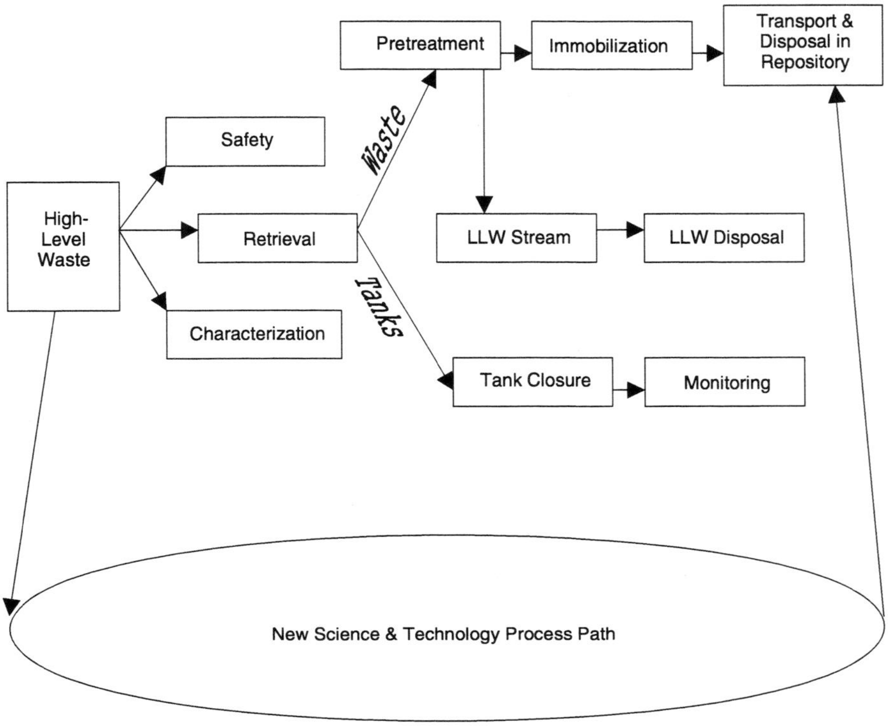

FIGURE 2.1 Generic waste remediation flow sheet used throughout this report. NOTE: LLW = low-level waste

they can be transported to a HLW repository.[1] Once the waste is removed from the tanks to the maximum extent that is technically and economically practical, the next step is *tank closure*. The retrieval of HLW must be adequate to ensure that residues can be declared as incidental to reprocessing and can be disposed of as LLW (see Sidebar 6.1). Tank closure consists of characterizing the residues, filling the tanks with cementitious material, and leaving the tanks in place. After closure, *monitoring* activities must take place to identify any release of residual waste into the environment.

HLW Generation and Cleanup Strategies at DOE Sites

The first HLW was generated at the Hanford Site from reprocessing reactor fuel for the production of plutonium used in nuclear weapons.

---

[1]The Yucca Mountain site in Nevada is currently under consideration as an HLW repository.

Subsequently, HLW was generated at three other sites, listed on p. 11. Most of the HLW was produced during World War II and the ensuing Cold War.[2] The generation of HLW and the cleanup strategy for each site are briefly described in the following sections.

The Hanford Site

The first recovery of plutonium at the Hanford Site began in late 1944 using the bismuth phosphate separation process (Gephart and Lundgren, 1998). This process recovered plutonium, but not uranium, from the spent fuel and produced large quantities of waste. Later efforts to recover uranium from the bismuth phosphate waste further changed the nature of HLW produced. The higher-activity liquid wastes from bismuth phosphate reprocessing were neutralized chemically to reduce their corrosiveness and stored in carbon steel underground tanks.

The first successful solvent extraction process for plutonium and uranium recovery in continuous plant operation was the REDOX (REDuction and OXidation) process, which began at Hanford in 1952. The REDOX process generated a lower volume of waste, which was chemically neutralized and stored in carbon steel underground tanks. In early 1956, a new solvent extraction process, PUREX (Plutonium and Uranium Recovery by Extraction), came into use at Hanford (Sidebar 2.1). PUREX used a different organic solvent and nitric acid. PUREX wastes were highly radioactive, self-boiling, and were also neutralized and stored in carbon steel underground tanks. Therefore, the tank wastes at Hanford include many different chemical compositions and physical characteristics, since different processes for plutonium recovery (as well as other operations) were developed and applied during the operational life of this site. Table 2.1 presents a simplified inventory of the wastes at the Hanford Site.[3] Only one percent of the mass of waste material is radioactive, but this is enough to make the tank contents highly dangerous.[4] Most of the radioactive species can be separated and immobilized for disposal as solid HLW; the residues remaining can be treated as other forms of waste.

---

[2]The HLW inventory may still grow slightly at the SRS because waste from deactivation and decommissioning operations is stored in HLW tanks. A small quantity of HLW is produced also from the reprocessing of part of damaged research reactor fuel and of the fuel "blanket" from the Experimental Breeder Reactor II. For more details on research reactor spent fuel, see (NRC, 1998).

[3]HLW from the other sites has a different and less complex composition compared to the Hanford waste.

[4]The radiation field inside the tanks can be as high as 10,000 rad per hour (100 gray per hour).

TABLE 2.1 Total Waste Inventory Based on Process Knowledge for All Hanford Tanks. Only some of the radionuclides present are listed, and the inventory does not take into account the decay in radioactivity since 1996. Although water is not listed, it represents a significant fraction of HLW (>200,000,000 kilograms). NOTE: 1 curie = 3.7 x 10^{10} becquerels.

Species	Total Inventory (10^3 kilograms)	Total activity (10^3 curies)	Half-Life (years)
Na^+	49,170		
Al^{2+}	8,372		
Fe^{3+}	1,872		
Cr^{3+}	994		
Bi^{3+}	492		
Zr as $ZrO(OH)_2$	214		
Pb^{2+}	280		
Ni^{2+}	183		
Ca^{2+}	620		
K^+	521		
TOC	1,818		
NO_3^-	14,093		
CO_3^{2-}	4,836		
PO_4^{3-}	3,795		
SO_4^{2-}	3,226		
Si as SiO_3^{2-}	662		
F^-	575		
Cl^-	1,099		
Sr-90/Y-90	0.44	123,203	29.12
Tc-99	1.94	3,356	213,000
Cs-137/Ba-137m	0.54	91,667	30
Np-237	0.20	0.141	2,140,000
Pu-239	0.77	4,696	24,130
Pu-Total	0.77	225.59	

NOTE: TOC = total organic carbon. The half-life of a radionuclide is the time after which half of atoms on average have disintegrated. SOURCE: Agnew et al. (1997).

In addition to the evolution of the waste characteristics, the designs of the waste storage tanks evolved as well. The earliest tanks were single-shell tanks (SSTs) of carbon steel. Reliance on a single shell has resulted in leakage of waste from some of the SSTs into the subsurface soil. Of the 177 tanks at Hanford, 67 SSTs are known or suspected to have leaked. The total quantity of leaked waste is estimated to be between 3.4 and 6.7 million liters containing between 0.45 and 1.8

million curies (1.7 x 10^{16} and 6.7 x 10^{16} becquerels), although there are large uncertainties in these estimates (Waite, 1991; ERDA, 1975; Agnew et al., 1997). The later double-shell tanks (DSTs) consisted of a first tank surrounded by a secondary containment tank. Figure 2.2 shows a diagram of a SST and a DST. The external shell of the DST provides an additional barrier to leakage of waste into the subsurface soil. There has been some leakage into the annulus space lying between the shell liners of some DSTs. All the free liquid from SSTs has been transferred to DSTs, thereby greatly diminishing the potential impact of future leaks. The risk from leakage in the future is expected to be associated almost entirely with retrieval of waste from the SSTs, depending on the retrieval method used. The vulnerability to leakage into the subsurface soil is an important reason for closure of the Hanford tanks. A summary of the key facts and figures about the tank wastes at the Hanford Site is presented in Table 2.2.

In its record of decision of 1997, DOE adopted a phased approach to tank waste management at the Hanford Site (DOE, 1997). Under Phase I, which will last approximately until 2018, 10 percent of the tank waste and 20 to 25 percent of the radioactivity is slated for retrieval and immobilization on a pilot scale for demonstration purposes. After validation of the processes, DOE will implement Phase II, the full-scale production phase, which will last approximately 30 years. Figure 2.3 shows DOE's baseline plan for HLW management in Hanford.

So far, the waste has been characterized to the extent feasible for its heterogeneous contents. Waste retrieval operations have just begun.

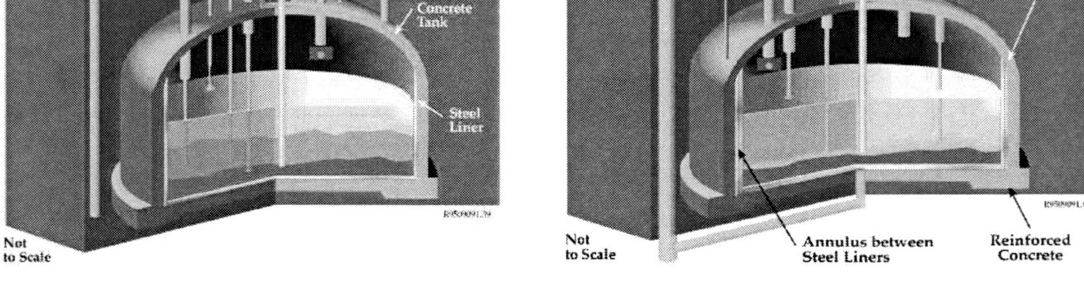

FIGURE 2.2 Diagram of a SST (left) and a DST (right) at the Hanford Site. SOURCE: DOE-Tanks Focus Area.

TABLE 2.2 Facts and Figures about the Hanford Site

Type of information	Data
Surface of the site	1,450 square kilometers (560 square miles)
Number of HLW tanks	177
Volume of HLW	208 million liters (54 million gallons)
Radioactivity	About 200 million curies (7.4×10^{18} becquerels)
Characteristic of HLW	Alkaline waste (pH approximately 12), saltcake, viscous liquid, sludge, and highly non homogeneous waste
Types of tanks	Carbon steel, SST and DST

Waste sluicing by pumping liquid to entrain the sludge is the technique chosen for retrieval of the waste from the tanks. According to the plan, after retrieval, the waste will be sent to a pretreatment facility to separate the solids, handled as HLW, from the liquid phase, which will be decontaminated further to remove the hazardous constituents.[5] The solids-liquid separation (SLS) process is followed by sludge leaching of the separated solids to dissolve non-radioactive components. Cesium, strontium, transuranic elements, and technetium (as necessary) will be removed from the liquid phase and sent to the HLW vitrification facility

[5]Further details on pretreatment and immobilization processes are given in Chapters 4 and 5.

FIGURE 2.3 Based on information gathered from DOE, this figure represents a simplified view of the baseline plan for waste management at the Hanford Site.

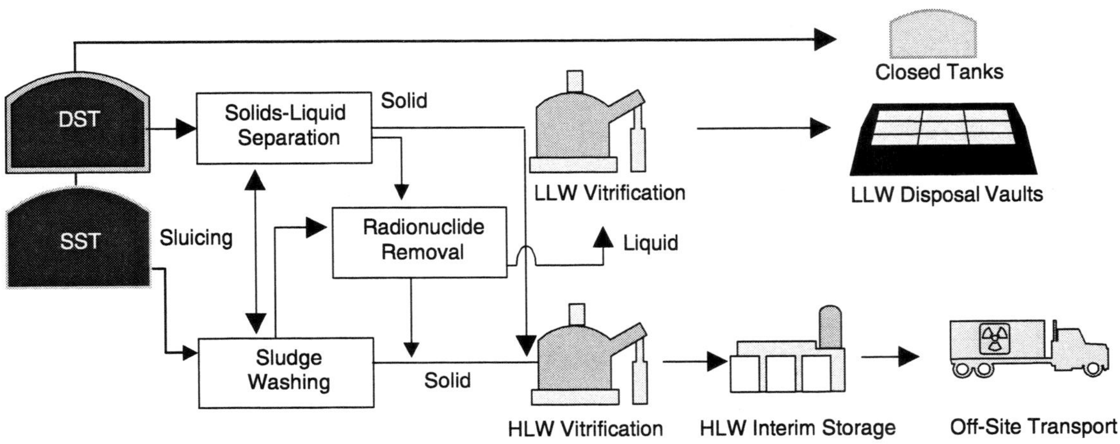

along with the HLW solids. The HLW glass canisters will be stored temporarily on site for eventual disposal in a geological repository. According to the 1994 revision of the Tri-Party Agreement (TPA) between DOE, the U.S. Environmental Protection Agency (EPA), and the Washington State Department of Ecology, LLW will also be immobilized in glass and stored on site (TPA, 1998).

The Idaho National Engineering and Environmental Laboratory

Between 1953 and 1992, the INEEL reprocessed spent nuclear fuel mainly for recovery of the fissile isotope uranium-235. In the reprocessing operation, spent nuclear fuel and its cladding materials (aluminum, zirconium, stainless steel, and graphite) were dissolved in highly acidic solutions (nitric acid, hydrofluoric acid, and sulfuric acid were used) and graphite was burned. INEEL's waste type, containing uranic and transuranic isotopes generated from the reprocessing of spent fuel, is unique within DOE HLW sites in that it was stored while still in highly acidic form.

Table 2.3 presents some general information on the wastes at the INEEL. The majority of INEEL's waste has been calcined to a granular solid (ceramic), which is considered an interim storage form by the State of Idaho. The calcine is stored in partially buried stainless steel bins (designed to last 500 years) grouped inside concrete vaults. Calcine waste can be processed further to convert it into a more consolidated long-term waste form. In addition, the INEEL has some liquid sodium-bearing waste (SBW) to be immobilized. This waste is mixed transuranic waste but, because of its radioactivity and large volume, it

TABLE 2.3 Facts and Figures about the INEEL

Type of information	Data
Surface of the site	2,300 square kilometers (890 square miles)
Number of HLW containers	11 tanks, 7 calcine vaults
Volume of HLW	About 5.3 million liters (1.4 million gallons) of liquid waste and 3.8 million liters (1 million gallons) of calcine
Radioactivity	520,000 curies of radioactivity (1.9×10^{16} becquerels) in liquid form and 24 million of curies (8.9×10^{17} becquerels) of radioactivity in calcined form
Characteristic of HLW	Very acidic (pH approximately 0), sludges and viscous liquid, calcine solids
Types of tanks	Stainless steel, single shell

is stored in underground tanks and managed as HLW (NRC, 1999b). DOE has recently decided (Huntoon, 2000) to seek direct vitrification as the preferred alternative for SBW. The preferred alternative for INEEL's calcine is to continue work on characterization, pretreatment, and vitrification options to be ready for shipment to a disposal facility by the end of 2035. Immobilized calcine and SBW will be stored temporarily on site awaiting eventual disposal in a geological repository.

Figure 2.4 shows the proposed baseline plans for the management of HLW calcine and liquid at INEEL. DOE is still studying this strategy and in 1999 was advised by the NRC on the different possibilities (NRC, 1999b).

The Savannah River Site

Plutonium and tritium for nuclear weapons were produced at the SRS near Aiken, South Carolina, by reprocessing spent fuel and targets. The PUREX process was used from the time the SRS opened in 1952; as in Hanford, acidic HLW generated by PUREX was neutralized for storage in carbon-steel tanks. SRS also produced plutonium-238 and uranium-233 and recovered them by processes other than standard PUREX. The

FIGURE 2.4 Simplified baseline plan for the management of liquid and calcined HLW at INEEL. Question marks represent alternative treatments for calcine. The final decision is still pending (Huntoon, 2000). For more details, see NRC (1999b).

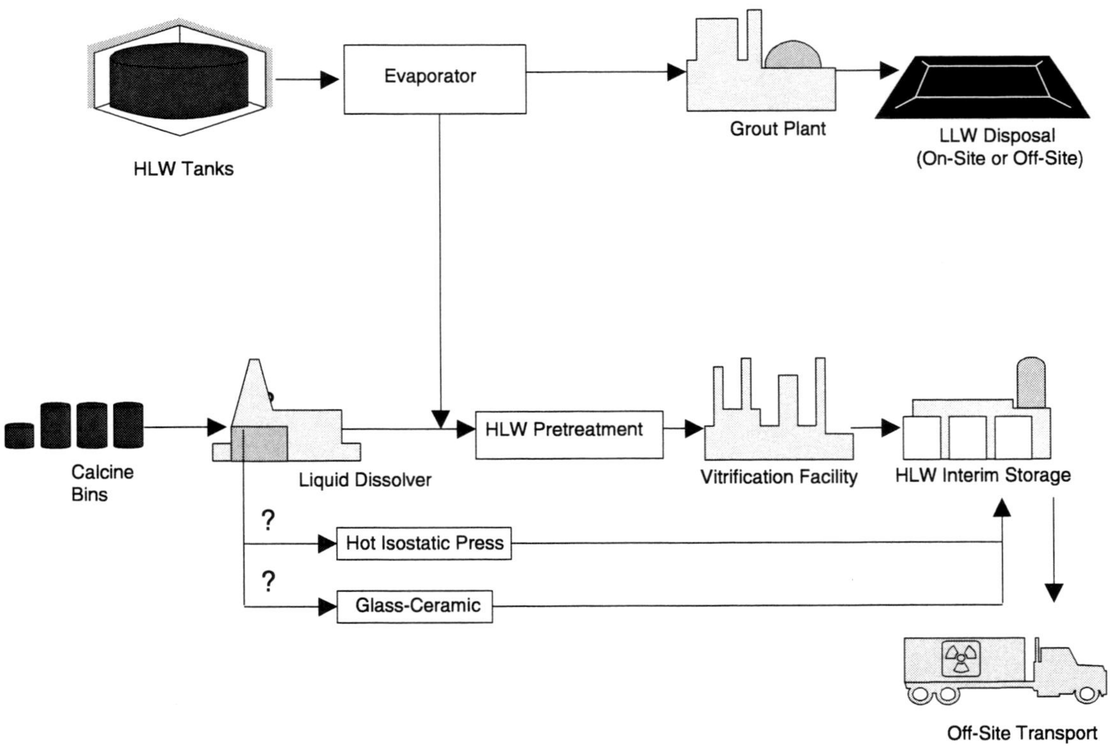

waste tanks contain HLW in the form of sludge, concentrated supernate, and saltcake. The sludge, of thin peanut butter-like consistency, contains 67 percent of the radioactivity and represents 9 volume percent; the supernate and the saltcake represent 91 volume percent and 33 percent of the radioactivity. HLW at the SRS is stored in 51 tanks, 2 of which have already been closed. Table 2.4 presents some of the principal data on the tank wastes at the SRS. Figure 2.5 shows the baseline plan for waste management at this site.

Radioactive vitrification operations at the SRS began in 1996, and approximately 1,000 HLW glass logs have been produced to date. The sludge is retrieved from different tanks and mixed in pretreatment batches to lower (blend down) the concentration of components that might add to the complexity of the vitrification step (more details are given in Chapter 5). The sludge is also pretreated by caustic extended sludge washing to remove aluminum from HLW. Aluminum is known to increase the viscosity of the molten glass, which complicates pouring the glass melt into the canisters and also considerably increases the total volume of immobilized HLW. The SRS is still in the process of selecting a preferred method to separate cesium, strontium, and actinides from the saltcake and supernate, after the previous process (large-scale in-tank precipitation) was abandoned because of technical difficulties (NRC, 2000b). An NRC panel is currently advising DOE on viable alternatives (NRC, 2001a). The leached sludge solids are then combined with cesium, strontium, technetium, and transuranic elements from the liquid decontamination step, and sent to the vitrification plant, the Defense Waste Processing Facility (DWPF). The decontaminated LLW filtrate is then grouted (i.e., immobilized by cementation). The HLW glass canisters produced are temporarily stored on site waiting for eventual disposal in a geological repository.

TABLE 2.4 Facts and Figures about the SRS

Type of Information	Data
Surface of the site	800 square kilometers (300 square miles)
Number of HLW tanks	51 (2 tanks already closed)
Volume	125 million liters (33.4 million gallons)
Radioactivity	240 million curies (8.9×10^{18} becquerels) in the sludge and 180 million curies (4.4×10^{19} becquerels) distributed between the supernate and the saltcake
Characteristic of HLW	Alkaline (pH approximately 14)
Types of tanks	The carbon steel tanks are differentiated into Type I, II, III, and IV tanks, according to their characteristics. Of the 51, only 27 tanks (type III) are double shell tanks

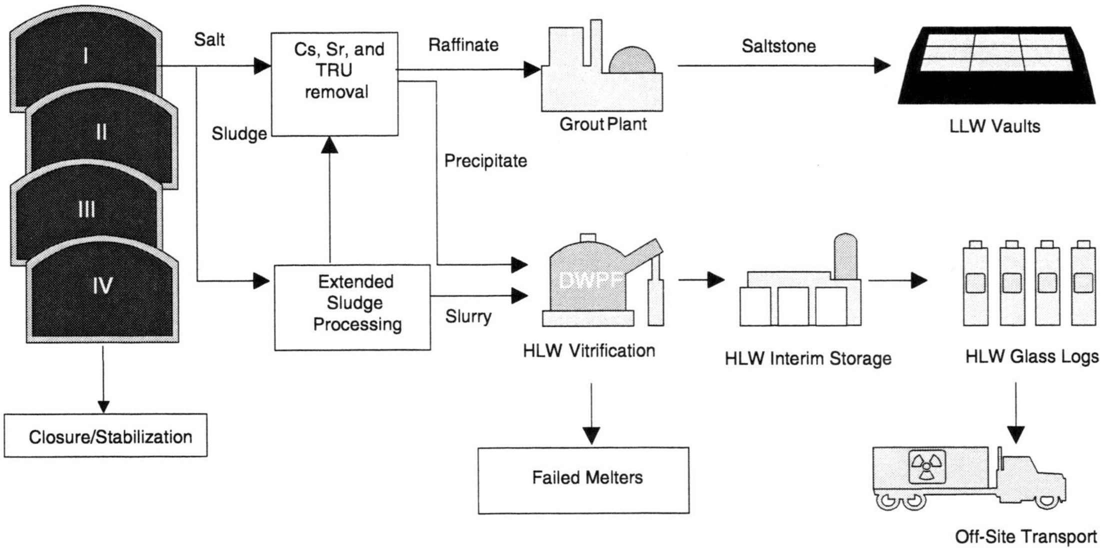

FIGURE 2.5 Simplified baseline plans for management of the waste at the SRS. The Defense Waste Processing Facility (DWPF) is the vitrification plant at the site. The waste supernate is the main source of cesium, strontium, and transuranic elements (Cs, Sr and TRU).

The West Valley Demonstration Project

The WVDP is situated near Buffalo, New York. The State of New York owns this site, not DOE. However, by the terms of the WVDP Act of 1980, DOE is responsible for retrieving the HLW on site, removing it for disposal, and decommissioning all parts of the site used for defense-related tasks. The U.S. Congress appropriates funds to DOE for this work, and New York State shares the costs. Because the WVDP has almost completed its HLW cleanup operations, the recommendations in this report are not as relevant to this site as they are for the Hanford, Savannah River, and INEEL sites. However, the committee gathered information on the WVDP because of the "operational experience" acquired at this site.

From 1966 to 1972, commercial spent nuclear fuel was reprocessed at the WVDP to recover uranium and plutonium by a licensed, commercial fuel reprocessing plant. The reprocessed reactor fuel included a substantial quantity of typical light water reactor (LWR) fuel, a single thorium-based core from an LWR, and a substantial amount of low-burnup defense spent fuel from the Hanford N Reactor, provided as a base load to support this first commercial reprocessing plant. The N Reactor fuel and the typical LWR fuel were reprocessed to recover uranium and plutonium using the PUREX process; the thorium-based core was reprocessed to recover uranium-233 and uranium-235.

The single tank farm at the WVDP contains two carbon steel tanks (each with a capacity of 2.4 million liters), one for storage of alkaline waste and one spare, along with two stainless steel tanks (each with a capacity of 110,000 liters), one for storage of the acid waste from the

TABLE 2.5 Facts and Figures about the WVDP

Type of Information	Data
Surface of the site	0.3 square miles (0.8 square kilometers)
Number of HLW tanks	2 carbon steel tanks of 2.4 million liters (660,000 gallons), 2 stainless steel tanks of 110,000 liters (30,000 gallons)
Volume of HLW (initial conditions)	2.4 million liters (0.66 million gallons), after blending a small amount of acid waste into the alkaline waste
Radioactivity	0.6 million curies (2.2×10^{16} becquerels)
Characteristic	Alkaline waste (pH approximately 14)
Types of tanks	Double shell, carbon steel

thorium-based core and one spare. Table 2.5 presents some of the principal data on the tank wastes at the WVDP. All of the waste was blended together in one of the larger storage tanks. All of the tanks were integrated into a recycling pretreatment system. As a result, there was a single waste composition to process.

Figure 2.6 shows a simplified baseline plan for the management of HLW at the WVDP. The supernatant liquid was removed and pretreated with zeolite, a granular aluminosilicate ion-exchange material, to remove the radioactive cesium and strontium, yielding a separate LLW salt solution. The sludge remaining at the bottom of the tanks was washed, and the liquids were treated with zeolite. The remaining HLW, consisting of the washed sludge and the zeolite used for pretreatment, was vitrified in borosilicate glass using a slurry-fed Joule melter of a design similar to that employed at the SRS. The LLW was further evapo-

FIGURE 2.6 Simplified baseline plans for the WVDP based on information provided by DOE.

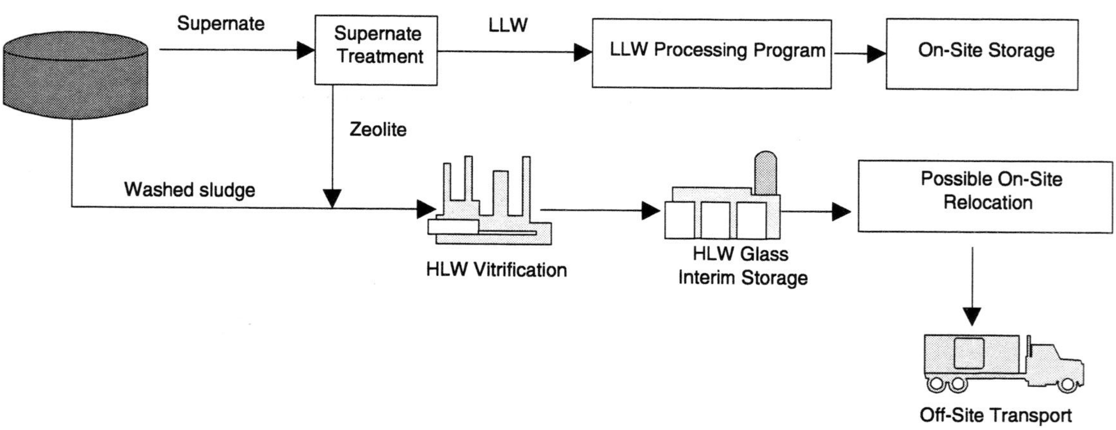

rated and blended with cement for interim storage on site. To date, the WVDP has almost completed[6] retrieval and vitrification of its HLW. The 255 glass canisters produced are stored temporarily on site, awaiting eventual disposal in a geological repository. The WVDP is now performing final flushing of the tanks to prepare for shutdown of the waste melter.

[6]More than 99 percent of the alpha emitters and more than 97 percent of the beta and gamma emitters have been retrieved to date.

3

Characterization

Characterization requirements relate to all parts of the HLW management process. Important parts of this process requiring characterization are the following:

- waste storage;
- retrieval;
- pretreatment;
- immobilization;
- tank closure;
- near-field monitoring; and
- state-of-the-equipment monitoring.

The successful performance of these HLW process activities depends to a great extent on characterization of the waste streams and of process equipment. Characterization requirements vary greatly depending on the relevant operational area but generally include measurements of chemical, physical, and radiological properties. The main challenges in characterization consist of difficulties in retrieving and manipulating samples of radioactive material, with the related risk of radiation exposure to workers; high costs; long turnaround times; and poor data reliability.

Poor data reliability occurs when only a limited number of samples is available and these samples are not representative of the parameter measured. For example, samples from HLW tanks are known to be unrepresentative of the total contents because of the heterogeneity of waste distribution. Moreover, the complicated procedures necessary during retrieval, transport, and measurement of highly radioactive samples significantly increase the chances of error.

Research with the objective of improving characterization operations will contribute to increased safety, efficiency, and cost-effectiveness.

Characterization Issues

The committee identified long-term basic research needs related to characterization activities throughout the HLW treatment process. **The objective of the recommended long-term basic research for characterization is to provide the scientific basis for developing innovative methods acquiring real-time and, when practical, in situ characterization data for HLW and process streams that could be useful for all phases of the waste management program.** Research needs for characterization are condensed into the two general ones listed at the end of this chapter. Analysis terms used in this report are defined in Sidebar 3.1. Table 3.1 provides examples of types of characterization activities sorted by process area.

Characterization for Waste Storage

High-level waste, whether stored in tanks as solids-liquid mixtures or in vaults as calcine, must be characterized to determine chemical (elemental and molecular), physical, and radiological properties to identify major safety concerns and to formulate subsequent treatment

SIDEBAR 3.1 DEFINITION OF ANALYTICAL TERMS

- **Sampling and analysis: Removal of a portion of the material to be characterized and determination of its properties, using various established methods. Given the state of current practices, sampling and analysis require a significant turnaround time (typically days to months).**
- **Turnaround time: Time between the beginning of a characterization process and the reporting of the data.**
- **In situ or on-line measurement: Measurement of a material's characteristics in place. The result can be displayed either at the site of the measurement or remotely. An example would be the determination of a chemical composition by spectroscopic analysis. For instance, the source of electromagnetic radiation could be supplied by a fiber-optic cable inserted into the region of interest. Results are usually available in near real time.**
- **Real time: Instantaneous turnaround of result data.**
- **Near real time: Turnaround time of a few seconds to a few minutes.**
- **Remote measurement: Determination of a material's characteristics from a distance without physical contact with the material. An example would be the determination of chemical composition by analysis of gamma radiation emitted after interrogation using neutrons from a remote source. Aerial or satellite fly-by analysis is an extreme example of remote measurement.**

TABLE 3.1 Desirable Types of Characterization and Examples of Analytes in Different Process Areas

Process Area	Type of Analysis	Examples of Analytes
Stored waste	Chemical	• Elements and molecular species in condensed and gaseous phases of waste and chemical speciation • pH
	Physical	• Same properties as for the retrieval and pretreatment process area • Foam or crust thickness in tanks
	Radiological	• Major radionuclides
Retrieval	Physical	• Density • Rheology • Percent solids • Height of the liquids in concentrated salt solutions and sludges
Pretreatment	Chemical	• Major glass-forming components such as B, Na, Ca, Al, and Si associated with melter feeds; these analyses currently determine the glass product acceptability • Metals associated with melter life and product quality, such as noble metals, Mn, Fe, Ni, and Cr in melter feeds • Anions associated with glass product quality and melter corrosion, such as sulfates, phosphates, fluorides, and chlorides • RCRA materials in the secondary waste streams • Organics that may affect the redox potential of the glass melt
	Physical	• Liquid levels in concentrated salt solutions and sludges • Particle size distribution • Density
	Radiological	• Radionuclides associated with high-activity fields and HLW such as Cs-137, Sr-90, U-235, Tc-99, Np-237, and other transuranic elements
Immobilization	Chemical	• Redox potential of the melter feed • pH • Organic materials that affect melter operation such as tri-N-butyl phosphate, multi-ring aromatics • Volatile species present in off-gas stream such as RCRA materials, NOx, and SOx; toxic gases; and Hg and other semivolatile metals such as B, Cs, and Na
	Physical	• Viscosity • Density • Foam or crust thickness in the melter
	Radiological	• Radionuclides in the off-gas stream, such as Tc, Cs, and I
Tank Closure	Chemical	• Metals associated with environmental concerns such as Pb, Hg, Cr, Np, and Pu
	Radiological	• Residual radionuclides remaining in tanks and process facilities, such as Cs, Np, Tc, and Pu

NOTE: RCRA = Resource Conservation and Recovery Act

steps. Determination of the presence of foam or crust layers in tanks is also important for safe storage, because such layers may trap hazardous gases, which are then released in "burps" causing potential explosion hazards. Data have been acquired for HLW stored in tanks at Hanford and SRS, but very few are available for INEEL calcine. Core, liquid "grab" (using the "bottle-on-a-string" method), and vapor-phase sampling are the methods currently used to characterize waste in the tanks. Sampling from tanks is very costly and has a lengthy turnaround time for analyses. The cost varies from thousands of dollars for limited analysis of liquid or gaseous samples to hundreds of thousands of dollars for core samples. Turnaround times vary from a few days for vapor-phase samples to typically 180 days for core and liquid grab samples, depending on the extent of characterization required and the difficulty of acquiring the sample. These long turnaround times are due in part to administrative procedures. In practice, the operator must develop sampling plans, obtain radiation safety approval, and set up the equipment to retrieve the sample and send it to the laboratory. However, most of the delays still relate to technological limitations in manipulating highly radioactive samples.

Characterization for Retrieval

Retrieval from tanks requires physical data on the properties of solid materials transport and chemical data on dissolution and compatibility of different waste streams. Pipeline transfers require data on physical properties, such as density, particle size, viscosity, and fraction of solids, primarily to facilitate pumping and to avoid plugging the pipelines (Table 3.1).

Characterization for Pretreatment

As HLW from multiple tanks is blended together for subsequent treatment, its composition changes, thus requiring new chemical, physical, and radiological characterization (Table 3.1). There is also a need to characterize feeds and process streams to avoid unwanted reactions and secondary products when blending. A formal review process takes place at Hanford for every waste transfer resulting in blending compositions to ensure that certain waste compatibility criteria are met. Characterization is an obvious key problem with regard to the quality of this review process. These analyses usually involve determination of the major glass-forming components such as boron, sodium, aluminum, and silicon. These compositions must be known to percentage levels. Minor components that are detrimental for the immobilization process (see Chapter 5), such as sulfate (SO_4^{2-}), phosphate (PO_4^{3-}), fluoride (F^-), and chromium (Cr^{3+}), must be measured to the tenths of a percent level. These analyses ideally are obtained from a well-blended batch of feed material of uniform and consistently known composition, over some significant interval of time.

Radiological characterization of cesium, strontium, and other transuranic elements during the pretreatment steps is necessary to verify that the requirements of radionuclide separations processes have been met. Residual levels of radioactive materials in LLW streams are in the order of microcurie-to-millicurie per gram (3.7×10^4 becquerel per gram and 3.7×10^7 becquerel per gram, respectively). Secondary system effluents from pretreatment must also be characterized for environmental purposes. The release of hazardous chemicals into the environment is controlled by the EPA. Some elements that must be quantified to comply with EPA's regulations are lead, mercury, and chromium.

Characterization for Immobilization

Vitrification in borosilicate glass is the current baseline method chosen by DOE for immobilizing HLW. Assurance that the waste glass is of sufficient quality to meet the waste acceptance criteria (see Sidebar 5.3 in Chapter 5) rests entirely on control of the vitrification feed composition and process conditions (for instance, feed flow rates and melter temperature). That is, instead of sampling the final glass product, the amounts of HLW and glass-forming materials (vitrification feed) are maintained within a predetermined composition envelope before they are melted to ensure that the product is in compliance with the desired end waste form. Once the glass log is made, there is no provision for analytical verification of its acceptability or for recovery of the final glass waste form if its properties are unacceptable. Therefore, chemical and physical characterization of the vitrification feed is an extremely important parameter in immobilization operations (for additional details, see Chapter 5).

An important factor in determining the rate of wasteglass production at the SRS and WVDP is the time required to analyze the melter feed and to verify that the correct amount of glass-forming material has been mixed in the waste stream. The current in-cell analytical methods involve mixing samples of the waste sludge and frit, melting, redissolving the resulting glass in hydrofluoric acid, and analyzing the sample by inductively coupled plasma-emission spectrometry. The entire process is very time consuming (typically 36 hours per sample) and, hence, very expensive.[1]

---

[1]The cost of analyzing a glass sample at the SRS is approximately $2,000 (Jantzen, Personal Communication, 2001).

Characterization of the glass stream as it is poured from the melter would provide significantly more information and shorter turnaround times. Glass stream characterization presents a challenging task, because the measurements would have to be performed in the presence of high background radiation fields. However, characterization of the glass stream would confirm that the glass falls within the envelope of acceptable compositions and would determine whether species insoluble in the glass, such as crystalline spinels and noble metals, are trapped in the melter or are leaving with the glass product. Therefore, on-line characterization of the glass stream is a highly desirable process control option. One method to achieve this goal might be to develop or adapt a remote analysis method based on the absorption or emission of electromagnetic radiation by the glass stream. Since the glass stream poured from the melter to the canister is hot and luminous, the absorption or emission of electromagnetic radiation could be analyzed and calibrated against different glass composition and/or glass properties.

Characterization for Tank Closure

Characterization for tank closure is a topic more fully discussed in Chapter 6, in a section with the same title; therefore, it is briefly summarized here. As the HLW is retrieved from the tanks, a fraction of radioactive solids may remain in the tanks as "crust" or dense "hard-heel" solids. Characterization of tanks, crusts or heels is necessary to determine their chemical and radionuclide (actinide content, alpha, gamma, and beta emitters) composition. Residual HLW left in the tanks must also be characterized to demonstrate that the waste has been adequately removed and that the residues can be considered "incidental waste" (see Sidebar 6.1) and be left in the tanks. Characterization research needs for tank closure are very similar to characterization needs for deactivation and decommissioning activities. These are addressed by a different NRC committee (NRC, 2000c, 2001b).

Characterization for Monitoring the Near Field

The area immediately around and below the tanks (the "near field") must be characterized to monitor potential waste leaks from the tanks. This issue is addressed in Chapter 6 as well. The quantity of waste that has leaked into the environment (i.e., soils) and the tank residuals are used as the collective source term in risk analyses for the site. The leaked wastes may contain hazardous chemicals such as lead, mercury, and chromium, as well as radionuclides. Knowledge of the chemical speciation of these hazardous materials in the tank, in the soil near the tanks, or in the LLW fraction left on site is crucial to track and control their subsequent migration and assess the threat posed. For instance, it is important to know the oxidation state of chromium, because its hexavalent state is more hazardous to the environment than its trivalent state. These species have to be measured at part-per-million (ppm) levels relative to the soils and groundwater. Activity levels for radionuclides in the environment are in the order of picocuries per gram (3.7×10^{-2} becquerels per gram) to microcuries per gram (3.7×10^{4} becquerels per gram). Characterization research needs for the EMSP in near-field monitoring are also addressed in a previous NRC report (NRC, 2000a).

Characterization of State of the Equipment

Characterization also includes determination of the state-of-the-process vessels and equipment to ascertain their continued viability in performing their operations. Examples of characterization of equipment properties that influence the HLW cleanup process are the following: tank integrity; stress corrosion cracking and pitting of melter electrodes, valve boxes, and pipelines; and the level of obstruction and location of plugs in transfer lines. Currently, evaluation of the process vessels is carried out by manual inspection techniques, if at all. These techniques are time consuming and may increase the radiation exposure of operating personnel. Therefore, knowledge of the operating state of equipment reduces the potential for personnel exposure to HLW and the planning required for maintenance intrusions in high-radiation areas. However, the equipment must be characterized frequently to ensure good operating conditions, to provide early warning of malfunctions, and (possibly) to increase public confidence in DOE's ability to operate the sites without generating additional environmental hazards. Finally, if process conditions could be adjusted in real time in response to actual equipment status, the likelihood and consequences of serious equipment or process failures could be reduced.

General Long-Term Basic Research Needs for Characterization

Long-term basic research is needed to provide the underlying principles for the characterization of chemical and physical properties of HLW, and to monitor the condition of the equipment. A long-term basic research program should contribute to development of the following:

1. *Remote-sensing instruments.* These instruments will reliably provide data on bulk properties or condition of the equipment, in a reasonable time frame (minutes to hours), at distances of tens of meters, in the order of picocuries per gram (3.7×10^{-2} becquerels per gram) for radionuclides and ppm levels for chemical compositions. Examples of current technologies that perform measurements at a distance in reasonable time frames are prompt gamma neutron activation analysis that measures elemental compositions; radiation monitors; infrared temperature indicators; and sonar measurements for wall thickness.
2. *On-line or in situ instruments.* These instruments will reliably provide data on bulk properties or condition of the equipment in real time or almost real time without retrieving samples, at ppm to percent levels for chemical compositions and in the order of picocuries per gram (3.7×10^{-2} becquerels per gram) for radionuclides. Examples of current technologies that perform measurements in real time are pH probes, radiation monitors, thermocouples and pressure gauges, electrolytic corrosion monitors, and fiber-optic spectrometers.

Of these techniques, remote monitoring is technically the more difficult. Because of the distance requirement, it will likely necessitate measurement of different properties than will on-line, contact measurements (see Sidebar 3.1). However, remote monitoring would allow more analyses on waste prior to retrieval, providing real statistics on waste composition, reduced radiation exposure to workers, and quicker turnaround times.

Desirable types of characterization and examples of analytes in different process areas are listed in Table 3.1. Examples of possible fields in which research is needed to develop improved methods for performing physical, chemical, and radiological characterization are the following:

- communication systems and signal processing;
- acoustic, microwave, and electromagnetic sciences for monitoring purposes;
- analytical chemistry using electromagnetic radiation for remote analysis;
- material sciences to identify radiation-resistant materials used for characterization purposes (probes, sensors, monitors, robots); and
- quantum physics and optical materials for optical communication and optical signal processing.

It is important to emphasize the need for developing long-term basic research for characterization purposes across the spectrum of the waste management process activities. Different DOE's focus areas involved in cleanup activities such as subsurface contamination and deactivation and decommissioning have similar characterization needs. Therefore, it is necessary to coordinate the characterization efforts among different programs and across the DOE complex.

4

Retrieval and Pretreatment

Retrieval of waste consists of removing the bulk of the waste from the underground tanks or bins and safely transporting it to the process facilities. Waste retrieval is treated in this chapter as part of the pretreatment process and will also be discussed in Chapter 6 with regard to tank closure issues. Pretreatment consists of separating the HLW components from the bulk of the waste and preparing the waste stream for feeding to immobilization processes. This preparation is necessary to remove constituents that interfere with the HLW immobilization process and achieve the final objective of reducing the volume of solidified HLW produced and meeting quality criteria.

Since most defense HLW was neutralized (with the exception of INEEL waste) to minimize tank corrosion, it contains large amounts of a wide variety of non-radioactive salts and metal hydroxides. Sidebar 4.1 describes the pretreatment programs for the HLW sites. The pretreatment of defense HLW is a much more complicated problem than the pretreatment of relatively pure acidic HLW from processing commercial reactor fuel (as performed in some foreign countries such as France, United Kingdom, and Japan) because HLW is not neutralized. The pretreatment required depends on the chemical and radiochemical composition of the waste. Because of the complex and varied chemistry of the elements present in the mixture, it is usually not possible to find a single reagent or chemical process efficient enough to accomplish all of the separations needed. Therefore, multistep processes are normally required. The unit operations that are performed as part of pretreatment (Figure 4.1) include the following:

- retrieval and blending;
- solids-liquid separation;
- sludge leaching; and
- liquid decontamination.

SIDEBAR 4.1 SITE-BY-SITE PRETREATMENT BASELINES

Pretreatment at the Hanford Site

Retrieval of waste at the Hanford Site has just begun; in fact, to date, none of the tanks has been completely emptied. After retrieval, the alkaline waste will be sent to a pretreatment facility to separate hazardous constituents from bulk material. The SLS process is followed by enhanced sludge washing. Cesium and strontium will be removed by ion exchange from the liquids and sent to the HLW immobilization facility along with the HLW solids. The Hanford Site has selected vitrification in borosilicate glass for immobilization of the LLW stream, as well as the HLW.

Pretreatment at INEEL

The proposed plan includes the "non-separations alternative early vitrification option," which encompasses (1) direct vitrification for the liquid SBW, with use of the current tank farm ending by the end of 2012, and (2) for the calcine, enhancement of characterization, retrieval, and treatment technology, leading to vitrification with or without separations by 2035. The method for processing and disposal for calcine will depend on the results of future separations research (Huntoon, 2000). A previous NRC committee on alternatives for HLW treatment at INEEL concluded that calcine is chemically stable and safely stored; therefore no immediate action should be taken (NRC, 1999b).

Pretreatment at SRS

The alkaline salt solutions retrieved from different tanks are mixed in pretreatment tanks to "blend down" components that might cause problems in the vitrification step. The sludge is separated and pretreated by caustic leaching to solubilize aluminum from the HLW feed. The SRS is still in the process of selecting a method to separate cesium from the supernate, dissolved saltcake and sludge-leach solutions. The previous method (simultaneous large-scale in-tank precipitation with tetraphenylborate to remove cesium, and sorption with hydrous sodium titanate to remove strontium) was discarded because of technical difficulties related to rapid release of benzene (NRC, 2000b). Studies to resolve the benzene problem, as well as to identify alternative processes, are currently under way. In the meantime, HLW sludge is being vitrified for future disposal (see Chapter 5). Pretreatment at SRS also includes adding formic acid to reduce mercuric compounds in the waste to metallic mercury. This mercury is then steam-stripped from the feed directed to the immobilization facility to avoid its presence in the melter.

Pretreatment at WVDP

After the sludge remaining at the bottom of the tanks was washed, the wash supernate and the original supernate liquid were pretreated with zeolite to remove the radioactive cesium, strontium, and residual transuranic elements. The decontaminated supernate was concentrated by evaporation and immobilized with cement. The high-level sludge and zeolite were combined and vitrified in borosilicate glass. The WVDP has retrieved and vitrified more than 95 percent of the initial waste. Zeolite was not selected for treating supernates at other sites because it would generate an excessive volume of vitrified HLW.

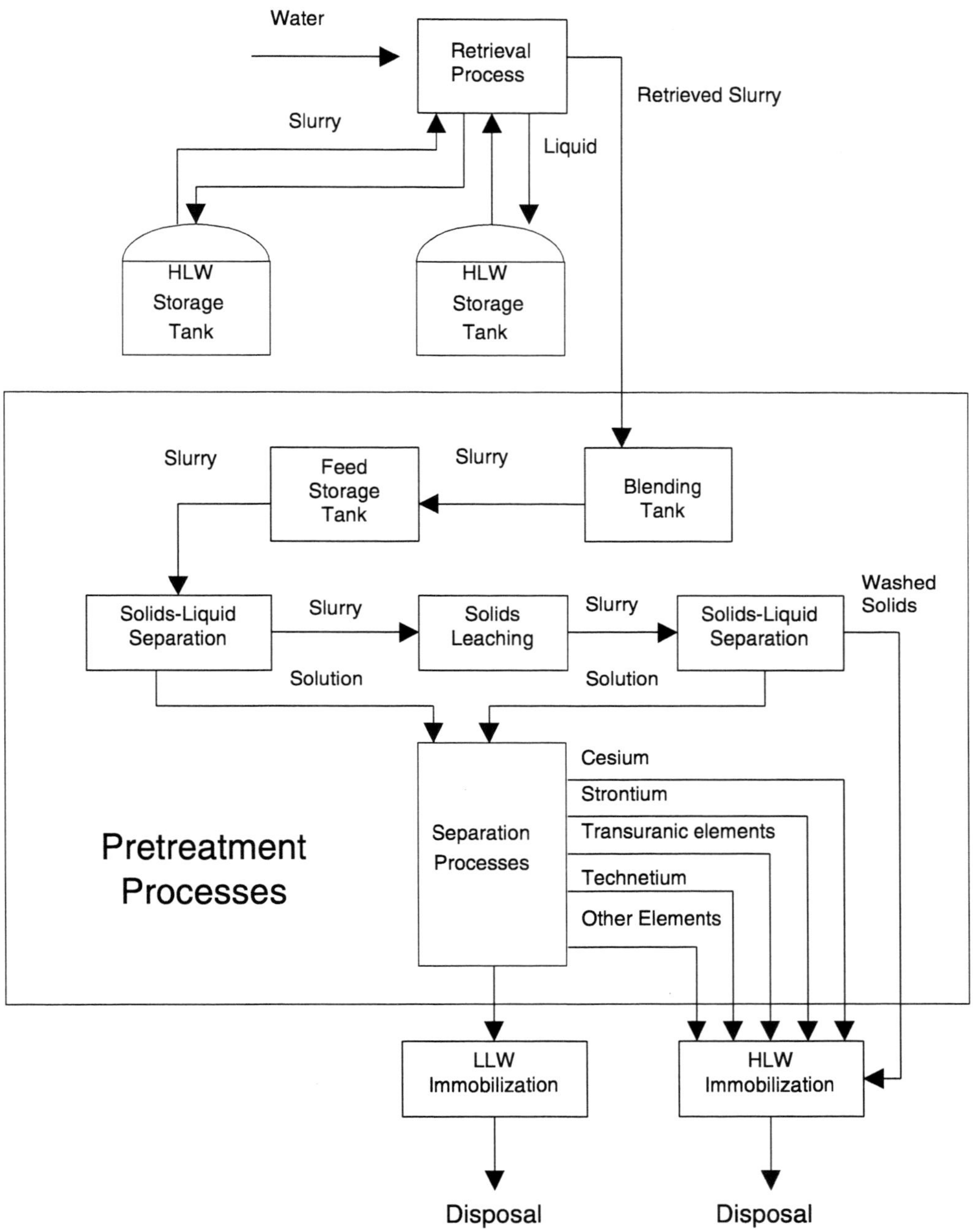

FIGURE 4.1 Simplified scheme of the pretreatment process.

The first step consists of the retrieval of waste slurry from the tanks and the mixing of various batches of feed materials to achieve a reasonable volume of feed with uniform properties. SLS is performed to separate solids from the slurry, leaving a solids-free solution. This separation may recur at several points in the process. Sludge leaching consists of the selective dissolution of specific constituents from the solids in the retrieved waste slurry. Liquid decontamination consists of the selective removal of radioactive and hazardous species from the liquid streams. The solids from sludge leaching and the species removed during the liquid decontamination step are combined to generate the feed for HLW immobilization. If key radionuclides have been adequately extracted, this waste is incidental to reprocessing, or WIR, and it is classified according to its composition (see p. 11). In addition to small concentrations of radioactive species, WIR may contain significant amounts of hazardous wastes (such as chromium, mercury, and lead) as well as large amounts of soluble salts (such as $NaNO_3$, NaOH, and $NaAlO_2$). Additional constraints are imposed if WIR contains hazardous material causing it to be classified as mixed waste, as defined in Sidebar 2.2. Disposal of WIR, although less expensive than that of HLW, still represents a high cost because of its large volume.

Pretreatment Issues

Pretreatment operational issues, long-term research needs, as well as general approaches used at the different sites are described below. **The objective of the long-term basic research recommendations for pretreatment is to provide the scientific basis for developing high-efficiency, high-throughput separation methods that could reduce HLW program costs over the next several decades.**

Retrieval and Blending

To date, waste has been completely retrieved from two tanks, both located at the SRS. The DOE plans to retrieve the remaining waste from the storage tanks with methods used in the past at the Hanford Site and SRS. Most of the waste will be retrieved from the tanks by pumping out the slurry of supernate and solids. In the case of INEEL waste, calcine will likely be retrieved from the storage bins by vacuuming. Stabilized water (water adjusted for pH and REDOX potential to minimize corrosion of the steel tank) will be added to the residual waste. Following mixing to dissolve soluble components (e.g., saltcake containing alkali metal salts including cesium) and suspend insoluble solids (e.g., hydroxides of iron, chromium, nickel, transuranic elements, alkaline earth sulfates, and phosphates) the slurry will be pumped out as before. Recovery of residual waste when the tank is nearly empty will be done by sluicing, which may present problems in the case of leaking tanks, depending on the extent of recovery required.

Wastes from multiple tanks can be blended in a waste receipt tank, as currently done at the SRS, but this may not be done at the Hanford Site because of the lack of free tank space. Instead, the Hanford waste will be treated on a tank-by-tank basis (see Chapter 5). Blending of the waste feeds can be used to reduce the final volume of immobilized waste, by diluting those components that (1) have low solubility in borosilicate glass, which limits the achievable waste loading, and (2) have elevated concentrations in some tanks. The "blended" components will therefore be less limiting with regard to waste loading. In general, the waste will reside in various tanks for an extended period prior to transfer for pretreatment, which will allow time for mixing, sampling, and analysis to ensure that it meets the feed specifications for the pretreatment steps, discussed below.

Long-Term Research Need

After some discussion, the committee decided that blending and retrieval is not a fertile topic for basic research because it would overlap with other research activities undertaken within the EM, such as those of the Tanks Focus Area (TFA) (DOE-TFA, 2000b).

Solids-Liquid Separation

Solids-liquid separation is required to physically separate the insoluble components from the supernate and leach solutions. This is a key operation because most of the hazardous material (e.g., strontium, transuranic elements, but not cesium) is associated with the solids. Since the clarified liquid must meet the stringent decontamination requirements for LLW, it is necessary to remove solids to an unusually high extent. For example, decontamination factors[1] (DFs) in the order of 10,000 to 100,000 may be required for some radionuclides to meet specifications for the immobilized LLW product; but solids removal by such large factors is not commonly achieved in practice in a single process cycle.

Solids-liquid separations will likely be required at more than one point during the process. Although simple settling and decantation will be adequate for some processes, it is necessary to remove the solids with an extremely high efficiency, at least once, during the process cycle to meet the decontamination requirements for all of the constituents in the clarified liquid. Inadequate SLS will impact downstream operations, for example, by causing plugging of ion-exchange columns

---

[1]Ratio of concentration of a species in the feed to that in the effluent of a process (see the glossary in Appendix G).

and by degrading DFs for most processes. This operation is critical to the operability and performance of pretreatment, and its success is not assured.

Long-Term Research Needs

Long-term basic research is needed on filter media material and on filtration methods using appropriate simulated slurries followed by verification using actual waste. The objective is to improve the rate of filtration, filter media life, resistance to plugging or fouling, and removal efficiencies for very small solid particles and colloids (removal factors at least 10^4 with a range of particle sizes down to 0.1 micrometer, and possibly colloidal). Additionally, SLS methods other than filtration (such as centrifugal separation or flocculation and settling) should be investigated.

Sludge Leaching

After being separated from the retrieved slurry, the sludges are leached by mixing them with a solution of sodium hydroxide to dissolve some of the bulk constituents. The purpose of leaching is to remove compounds that would be detrimental to the immobilization process or to the quality of the immobilized product. Examples of such compounds are sodium and aluminum, which are large contributors to the bulk waste material, and chromium, sulfate, and phosphate, which, although present in much smaller amounts, are relatively insoluble in borosilicate glass and interfere with vitrification operations. If not removed, these non-radioactive materials would increase the final volume of the HLW and/or lead to a hard-to-process slurry feed or an unacceptable final product.

Removal of these undesirable species from the sludges is currently based on the "enhanced sludge leaching" method that is being used at the SRS and will likely be applied at the Hanford Site. This technique consists of leaching sludges with strongly alkaline solutions with the purposes of (1) solubilize aluminum and some other elements and (2) metathesize[2] and thereby partially solubilize anions such as phosphate and sulfate from insoluble salts (e.g., $CaSO_4$). The main problem with the enhanced sludge leaching method is that a large fraction of the waste (hydroxides of iron, manganese, nickel, zirconium, and other metals) cannot be dissolved in alkaline conditions. An alternative approach is to leach with acidic solutions to dissolve more of the solids, thereby leaving very little sludge. This was considered for

---

[2]Metathesis is a double-decomposition chemical reaction of the type $AB + CD \rightarrow AD + CB$ that is driven by the law of mass action when there is an excess of one ion.

Hanford waste (Swanson, 1993) and is still under consideration for INEEL waste, because alkaline leaching is not effective for the zirconium-type calcine (NRC, 1999b). In fact, except for limited cases involving complexing with concentrated fluoride or oxalate, zirconium can be maintained in solution only under acidic conditions. Zirconium is the dominant element that has to be separated only for INEEL waste; however, in all cases, the leach solutions require a subsequent SLS step.

Operations using precipitation and solids leaching can be affected negatively by several factors:

- the presence of high salt concentrations and organic complexants that may solubilize strontium, and perhaps transuranic elements;
- the solubility and speciation of actinides (plutonium, neptunium, americium, uranium) and strontium in caustic media containing carbonates and organics (e.g., oxalic acid); and
- secondary reactions in leach solutions that can lead to inadvertent reprecipitation and gel formation. These reactions have been observed in some aged or mixed Hanford leach solutions that were saturated in many components.

A number of unknown reactions could occur during solids leaching and washing, with uncertain consequences for the chemical and physical properties of the waste stream. Slow reactions can occur between aluminates, silicates, and other materials to form complex solids. Leach solutions and clarified supernates may be mixed together and/or evaporated, thereby changing their composition. Since leach solutions were initially saturated in several components, it is not surprising that new insoluble compounds can be generated. This has two main implications for the processing: (1) SLS processes must be performed several times, and (2) solids can continue to precipitate, which can disrupt operations such as filtration, ion exchange, and evaporation.

A large amount of research has been done on solubilities in multicomponent systems, for example by Rapko and Lumetta (2000), and efforts have been made to model the results for application to the extremely complex waste systems. At Hanford, for example, a thermodynamic equilibrium program called Environmental Simulation Program® is used for several purposes, such as estimating the consequences of waste mixing and dilution during transfers and retrieval (Mahoney et al., 2000; Papp, 1998). Extension of such models to include kinetic data and solid phase identification would be beneficial. At INEEL, there are only very limited data on dissolution of calcine in acid solutions, and more studies are needed if the waste is to be processed before immobilization (NRC, 1999b).

Long-Term Research Needs

Sludge leaching offers several long-term basic research opportunities for successful operation of the pretreatment processes for all of the sites. Projects should focus on chemical kinetic data (rates of dissolution) and equilibrium data (residual concentrations and solubility products) in very complex multicomponent systems. Examples of such components are: sodium, aluminum, silicon, iron, chromium, zirconium, sulfate, phosphate, carbonate, hydroxide, nitrite, and nitrate, many of them dissolved at saturation limits. In spite of the complexity of the task, research in this direction can yield useful results in the long term. Research is also needed on solution stability and precipitation mechanisms in highly concentrated solutions. Of particular interest is the stability of solutions with time, relative to reprecipitation of new phases and gel formation. To the extent practical, effects of radiolysis should be included in experiment with simulants.

Long-term basic research is needed to develop methods to enhance removal or to cope with the presence of problematic material and constituents not readily dissolved by alkaline leaching (e.g., decomposition of organic materials and dissolution of Cr^{3+} by oxidation to Cr^{6+}). In addition, the option of acid leaching of the sludge should be investigated to dissolve a greater fraction of the solids. Finally, research leading to the development of a predictive model for solids-liquid behavior in concentrated salt solutions could result in reduced costs and time.

Liquid Decontamination

After sludge leaching and SLS, the supernate and all leach solutions are carried through one or more processes designed to remove specific hazardous or radioactive constituents so that the liquid product can meet the acceptance criteria for LLW immobilization and disposal. As indicated in Figure 4.1, these processes may include removal of radioactive materials such as cesium, strontium, transuranic elements, technetium, and other elements, as well as non-radioactive toxic metals and organics. Several processes are known for each of these separations, such as zeolite ion exchange to remove cesium and strontium and hydrous sodium titanate ion exchange to remove strontium and transuranic elements. Since none of the processes is as simple and effective as might be desired,[3] there is continuing interest in finding improved separation methods.

---

[3]For instance, the monosodium titanate process employed at the SRS may not be successful for all waste streams since it may require extensive blending because of low DFs.

There is potential advantage in combining two or more separation processes into a single unit operation, for example, by combining processes for both strontium and transuranic elements removal, as well as for finding a single reagent that effectively will remove more than one problem constituent. An example of a combined process is the use of a mixture of chlorinated cobalt dicarbollide, polyethylene glycol and octyl(phenyl)-N,N-diisobutylcarbamoyl methyl phosphine oxide (CMPO), which has been used in Russia for simultaneous removal of cesium, strontium, and transuranic elements and has received limited testing at INEEL (Thompson, 1998).

Absorption (e.g., using activated carbon), coprecipitation, and ion exchange provide selective removal of contaminants, either on particles in columns or on absorbent solids that are added to the solution and then removed by SLS. The main issues to be addressed with absorption methods are the maximum achievable decontamination effectiveness and the amount of sorbent required to achieve such effectiveness. The loaded sorbent must either be routed to the HLW for disposal or be eluted and reused for subsequent loading cycles. If reused, the attainment of sufficiently large DFs (in the range up to 10^5) in subsequent cycles is usually very difficult. Solvent extraction is another option: a solvent extraction process using a crown ether is under study for SRS to remove cesium from the HLW liquid (NRC, 2000b).

Solvent extraction may be particularly advantageous for acidic solutions. Because of the unique composition of the INEEL HLW, the caustic leaching process used at the other sites offers little benefit. Acid leaching and separation of inactive components from the acidic solution is an option. The final immobilized HLW volume could be decreased substantially if acid leaching and acid-side processing (Swanson, 1993) were selected for either Hanford Phase II or INEEL calcine. The value of this approach depends on the availability of repository space for DOE HLW. Current plans call for HLW from Hanford and SRS to go to a first geological repository. HLW from INEEL will be directed to a second repository, particularly if processing is not completed before the closure of the first repository (NRC, 1999b, page 83). However, at present the United States is investigating only one site (Yucca Mountain) to determine its suitability to host a geological repository, and a decision to proceed has not yet been made. The possibility of a second repository is highly conjectural at this time.

In all cases, each separation process generates some quantity of secondary waste, such as ion-exchange sorbents, organic solvents, or various conditioning solutions. Dealing with these secondary waste streams is sometimes difficult because they can interfere with the separation processes and increase final immobilized waste volume. Therefore, it is necessary to develop processes that minimize secondary wastes or generate secondary wastes that are relatively innocuous. The strategies for pretreatment and for both HLW and LLW immobilization and disposal are mutually interdependent. Because of this connection, the nature of both immobilized wastes will be determined by the separation and pretreatment strategy adopted.

An important research effort in the pretreatment area is already carried out within the Efficient Separations Program-Integrated Program (ESPIP), a crosscutting project within the EM (DOE-ESPIP, 2001). Its mission is to identify, develop, and perfect separation technologies to separate cesium, strontium, and transuranic elements from radioactive waste streams. Additionally, the EPA has developed a large database of technologies that might be applicable for the removal of non-radioactive toxic materials (metals and organics from the HLW and LLW). The EPA Superfund Innovative Technology Evaluation (SITE) program evaluates all available information on the technology for hazardous waste remediation and analyzes its overall applicability to other site characteristics, waste types, and waste matrices. The objective of the SITE program is to encourage the development and implementation of (1) innovative treatment technologies and (2) monitoring and measurement. For further information see EPA-SITE (2001).

Long-Term Research Needs

A long-term basic research effort is needed to complement the activities within the ESPIP to identify sorbents and separation methods for cesium, strontium, technetium, and transuranic elements that are effective and operable in high-ionic-strength alkaline or acidic solutions and high radiation fields. Sorbents must have high selectivity and capacity and must either be capable of elution and regeneration for a number of cycles or be easily decomposed, such that disposal of the sorbent has a minimal impact on the final volume of immobilized waste. For example, inorganic sorbents should be investigated if their selectivity is sufficiently high (e.g., monosodium titanate for strontium and actinides, ammonium molybdato-phosphate for cesium).

Decontamination methods that combine removal of more than one target constituent within a single unit operation (e.g., a combination of sorbents or extractants that can remove cesium, strontium, and transuranic elements simultaneously from a feed composition) are needed and warrant further basic research. Research in solvent extraction separation methods is needed to identify stable, selective, high-capacity extractants that also are inexpensive and commercially available. In addition, research on solvent extraction methods would be particularly valuable for use with acidic waste solutions for which solid sorbents are often less effective. Solvent extraction from acidic solutions is the standard for reactor fuel reprocessing, because fuel is soluble only in acidic solutions. The waste treatment methods now being used are generally alkaline, mainly because the feed is usually alkaline and there is reluctance to add large quantities of nitric acid to re-acidify it. However, investigators should not restrict consideration to alkaline methods only, even though that is the starting medium. There could be benefit from treating solutions from acid leaching of alkaline sludge or INEEL calcine, but the methods would be very different from fuel reprocessing.

5

Immobilization

To reduce the risk of radionuclide transport to the environment, high-level radioactive waste must be immobilized (i.e., converted to a stable solid form) before its disposal in a geological repository. The durability of the immobilized waste form, together with the corrosion-resistant overpack protection against groundwater intrusion included in the design of the future repository, prevent or minimize migration of radioactive elements to the environment. Because repository space is limited, it is important to achieve the highest waste-volume reduction to minimize the number of containers needed for the immobilized waste. Hence, this would lead to reduced costs by more efficient use of waste immobilization, storage and transport facilities, and possibly to more efficient use of repository space.[1]

The current immobilization technique used by DOE for HLW is vitrification in a borosilicate glass matrix (see Sidebar 5.1). The DOE vitrification process involves blending HLW with borosilicate glass frit or glass precursor chemicals, such as oxides and carbonates. The mixture is heated in a Joule-heated melter (see Sidebar 5.2) to form a molten glass, which is then poured into stainless steel canisters and allowed to cool. The DOE is currently operating HLW vitrification plants at the SRS and at the WVDP. The current plan at the Hanford Site is to vitrify both HLW and LLW in borosilicate glass.[2] Finally, the INEEL will also immobilize its SBW in borosilicate glass and is now considering the option of vitrification for its calcined waste.

---

[1]The cost of producing an HLW glass log is approximately $1 million. A 1 percent increase in waste loading at the SRS could reduce cleanup costs by $200 million (Hrma et al., 1998).

[2]The TPA decision to vitrify the LLW at Hanford is a departure from the strategy used at the SRS and at the WVDP, where LLW is instead immobilized in a cement-based material, referred to as "saltstone" at the SRS (TPA, 1998).

SIDEBAR 5.1 SITE-BY-SITE HLW IMMOBILIZATION BASELINES

Immobilization at the Hanford Site

Both the HLW and the LLW will be vitrified at Hanford using Joule-heated melters, commencing in 2007. The Hanford tanks contain a wide range of reprocessing chemicals and wastes from the early bismuth phosphate and REDOX processes, as well as PUREX wastes. During Phase I of waste immobilization at Hanford, ending in 2018, approximately 10 percent of the HLW is to be retrieved and vitrified. Little or no blending between tanks to smooth any waste composition variability is planned during Phase I because of high costs and limited availability of free tank space. Furthermore, although the Hanford melter will also use Joule-heating to immobilize HLW in borosilicate glass, it is proposed to use a melter feed containing raw chemicals (oxides, carbonates, etc.), rather than frit, to give greater flexibility in glass-batch preparation.

Immobilization at INEEL

The site has calcined liquid, acidic HLW to produce approximately 3.8 million liters of granular solid waste that is currently being stored, pending a decision on final immobilization and/or disposal. The remaining 5.3 million liters of liquid SBW is to be vitrified (Huntoon, 2000). A previous NRC committee on the INEEL recommended in its report that the calcined material be stored until the repository location and waste form acceptance criteria have been established (NRC, 1999b). That committee also advocated further investigation of a number of viable candidate waste forms for the calcined wastes, in addition to borosilicate glass, with the main objective being to increase the waste loading. These candidate waste forms include (1) alternative glass compositions, including high-waste-loading glasses prepared in single-use containers; (2) crystalline ceramics prepared by hot uniaxial or isostatic pressing; (3) glass-ceramic materials; and (4) cement-based waste forms.

Immobilization at SRS

In the DWPF vitrification facility at the SRS, HLW is immobilized in a borosilicate glass matrix. The DWPF melter, described in Sidebar 5.2, uses a wet feed (approximately 50 percent water) comprised of a slurry of waste and frit. The waste originates from a two-year homogeneous batch where HLW, retrieved from different tanks, is blended. A target "window" of feed compositions, consisting of a ternary diagram based on mixtures of two waste feeds and a glass frit (Figure 5.1) is used to determine the composition of the melter feed. This window has been established on the basis of previous melting trials with slurry feeds of frit plus simulated wastes. In this process control strategy, called Product Composition Control System (PCCS), portions of the two-year batch of waste are fed into a tank where, depending on the waste feed characteristics, the amount of frit (of a given composition) is adjusted so that the frit volume is minimized and the predicted properties of the final glass will fall within the target window. A statistically designed variability study comprised of approximately 30 glass melts is performed on every waste batch (about every 2 years) to ensure that waste-frit mixtures are correctly predicted by the PCCS models. The use of this process control strategy and a large homogeneous two-year sludge batch minimizes the number of actual radioactive glasses that need to be analyzed from the canisters produced. Therefore, control of the final radioactive glass composition mainly relies on the PCCS process control strategy by ensuring that the melter feed composition is such that the resulting glass exceeds, with a 95 percent confidence level, the measured durability of the benchmark Environmental Assessment (EA) glass.

Immobilization at WVDP

Immobilization operations at the WVDP involve a Joule-heated melter similar to that of the DWPF. However, the process control strategy is different from that used at the SRS. The WVDP uses glass forming chemicals rather than premelted frit. During production, numerous analyses are made on large volumes of wastes, to allow the feed composition to be adjusted to fit within the target composition window. Numerous samples on every canister of the final glass product are also required to characterize glass-waste properties.

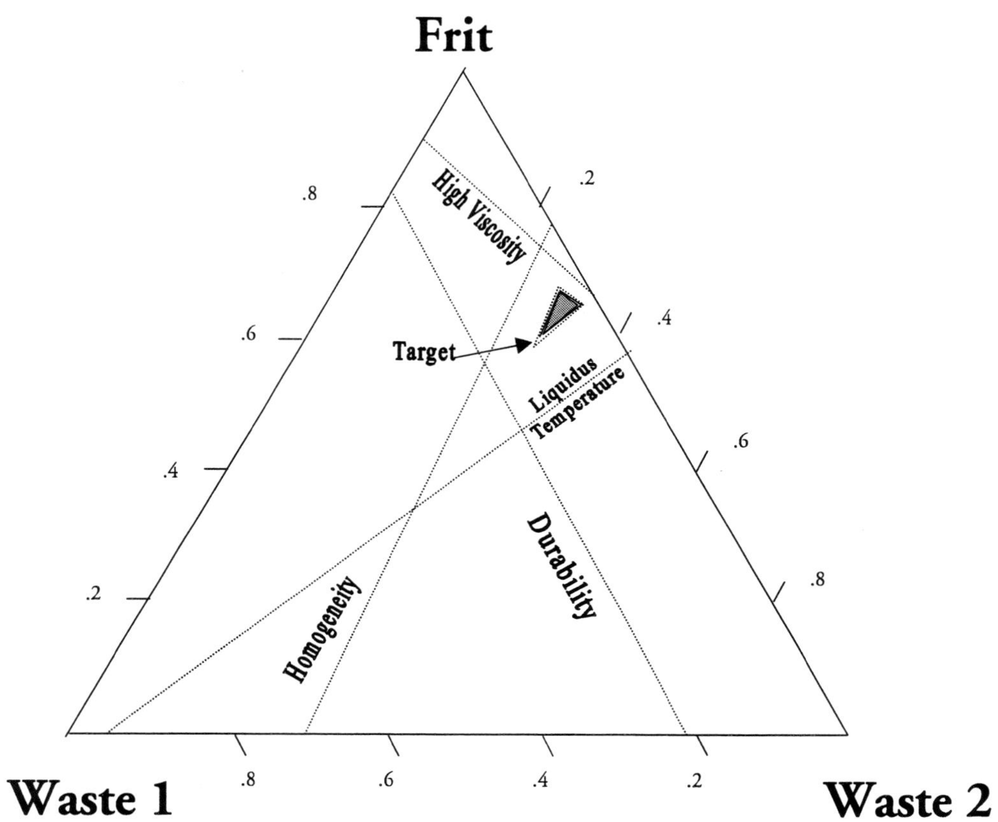

FIGURE 5.1 The PCCS window for the DWPF melter feed, based on a mixture of two types of HLW streams, Waste1 and Waste2, and glass frit. The solid white area in the diagram, called "Process Acceptable Region (PAR)," is defined by the mechanistic models in PCCS that bounds the range of allowable waste compositions and frit blends. The "target" point indicates maximum waste loading with the minimum amount of frit for a given batch of waste being fed to the melter. Waste-frit blends that fall within the PAR yield a final glass product that meets the appropriate processing and durability criteria. This PCCS allows minimal sampling on the final glass product while maintaining at least 95 percent confidence that the glass product falls within the qualified glass region. The window is defined in terms of relevant glass properties, including such constraints as melt viscosity, product durability and homogeneity, and the temperature for incipient crystallization (the liquidus). The viscosity and liquidus must be low enough to process the feed in the melter and pour it into the stainless steel canisters. The PCCS models that define the PAR target window have been established on the basis of

data from 300 to 400 non-radioactive and radioactive laboratory melts and pilot-scale melting trials. To verify the PCCS model and the DWPF process composition window, an additional 485 validation glasses, including 237 full-scale canister glass samples, taken during DWPF non-radioactive startup, were analyzed (Jantzen et al., 1998). The PCCS models were developed from glasses whose compositions cover the range of waste streams expected to be processed over the lifetime of the DWPF (Postles and Brown, 1991). The cost of establishing the DWPF process control system was approximately $5 million (Janzten, personal communication). SOURCE: DOE-Savannah River Site.

SIDEBAR 5.2 OPERATION OF THE DWPF JOULE-HEATED MELTER

The DWPF facility started non-radioactive operations in 1994 and was used to test a wide range of simulated wastes that covered the range of all wastes anticipated to be processed during the DWPF lifetime. Radioactive vitrification operations began at the DWPF in 1996. The Joule-heated melter in use at the SRS, shown below, is operated as follows. An initial charge of dry glass-forming ("batch") materials is fed to the melter and preheated by natural gas burners or electric heaters above the glass pool ("overhead plenum heaters"). The batch becomes sufficiently electrically conductive between 600 °C and 700 °C to allow further heating by electric current. The batch is then heated resistively to approximately 1150 °C by passage of an alternating current through the submerged melter electrodes (only 2 of the 4 electrodes are shown). The molten glass produced flows through a narrow region and is removed continuously from a side channel in the melter. The melter is replenished by feeding additional batch material onto the melt surface, where it forms a thick crust of unreacted material (the "cold cap") that serves to trap condensable volatile emissions and recycle them into the melt. The molten glass is then poured through the pour spout into stainless steel canisters and allowed to cool. Successful, continuous, long-term glass production normally requires a constant batch composition, uniform feed rate, and steady-state operating conditions.

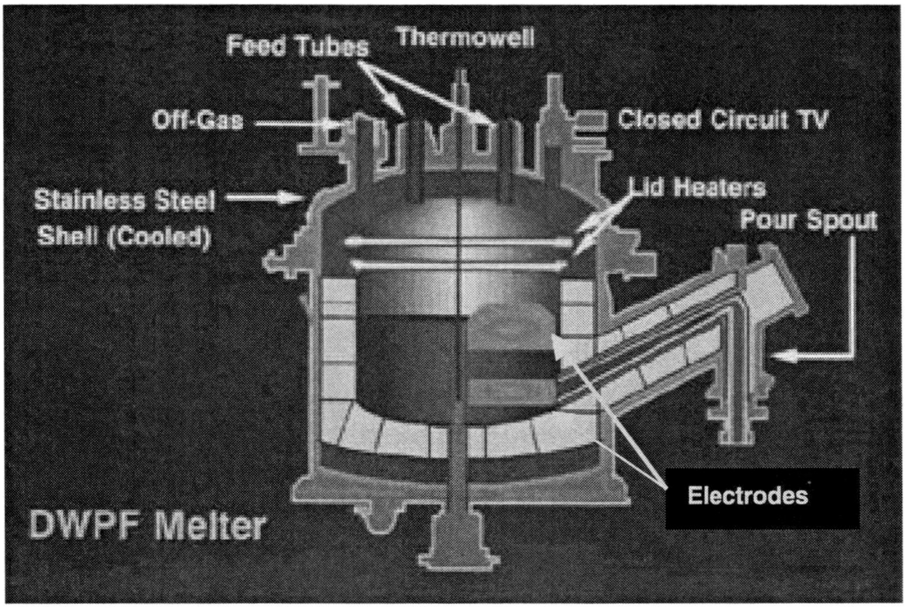

SOURCE: DOE-Savannah River Site.

The committee identified the following issues related to the choice of borosilicate glass and Joule-heated melters for the immobilization of HLW:

- limitations of borosilicate glasses as immobilization medium;
- crystal content of the borosilicate glass matrix;
- long-term leaching properties of the borosilicate glass matrix;
- use of unreacted glass-forming chemicals versus premelted glass frit;
- foaming in Joule-heated melters;
- precipitation of noble metals and crystalline phases in Joule-heated melters;
- limitations of Joule-heated melters in achieving higher processing temperatures; and
- alternative immobilization processes to Joule-heated melting.

Immobilization Issues

The issues listed above are described in the following section, along with research activities that could contribute to their resolution. **The overall objective of the long-term basic research recommended for immobilization is to provide the scientific basis to develop robust, high-loading[3] immobilization methods and materials that could provide enhancements or alternatives to the current immobilization strategy.**

Limitations of Borosilicate Glasses as Immobilization Medium

Under the current waste acceptance guidelines for the future geological repository (see Sidebar 5.3) the borosilicate glass waste form must meet certain performance requirements, developed on a case-by-case basis. Borosilicate glasses are appropriate immobilization media for many DOE HLW streams at the present level of waste loading (currently approximately 28 weight percent at DWPF on a dry calcine basis including Na_2O and SiO_2). However, it may not be possible to achieve substantial increases in waste loading using borosilicate glasses. Furthermore, these glasses may not be the optimum media for some problematic wastes such as the INEEL calcines, as

[3]Waste loading is the fraction of waste contained in a glass log or other waste form.

SIDEBAR 5.3 WASTE ACCEPTANCE CRITERIA

The Waste Acceptance System Requirements Document "determines the conditions necessary to be met by spent nuclear fuel and HLW, in order for DOE to be able to accept it for disposal" in a geological repository (DOE-OCRWM, 1999). The Yucca Mountain site in Nevada is currently under consideration for that repository. The DOE-OCRWM, which is responsible for the development of the HLW repository project, while the USNRC regulates the repository site, is currently developing the waste acceptance criteria.

Before 1977, the United States expected to reprocess all spent fuel from commercial reactors. It was intended that all HLW from reprocessing of commercial fuel would be immobilized by incorporation in a borosilicate glass matrix, prior to disposal in a geological repository. Glass-technology programs were initiated to identify suitable waste-glass compositions that would be resistant to leaching under many repository conditions. The glass waste form would thus constitute a primary barrier against release of radionuclides to the environment. A similar search for a practical glass waste form was underway in France and the United Kingdom, two countries that intended to reprocess their spent reactor fuel. When the United States decided, at the beginning of 1977, to forego reprocessing of commercial reactor fuel, the primary form of HLW became spent fuel, and the plans and regulations governing repository disposal were changed accordingly. Later, the DOE was authorized to dispose of HLW from the defense program sites in the first HLW repository. The defense wastes are expected to constitute only about 10 percent of the HLW in the first repository, with the balance of about 90 percent being spent fuel from commercial nuclear reactors. The initial defense HLW designated for the first repository will be borosilicate glass from the SRS and from the WVDP.

At the present time, the DOE wastes are scheduled for disposal along with spent commercial fuel in a common repository, where the performance acceptance criteria are based on spent fuel characteristics. The underlying objective for the DOE wastes is that disposal performance is predictable and at least as good as spent fuel. As part of the current SRS criteria for waste form acceptance, the borosilicate glass waste form must meet the following requirements (Janzten, 1993a; 1993b; 1999):

1. The glass must have a leach resistance greater than the EA glass, the benchmark waste glass identified in the DWPF EA.
2. The glass must exhibit no evident glass-in-glass phase separation.
3. The glass must be essentially free of crystal content.

These criteria are SRS guidelines for their current glass waste form, and do not apply to waste forms to be produced at the other DOE sites. The other sites might consider alternative waste forms, and would have to establish their own criteria to achieve comparable performance and acceptance.

stated by a previous NRC committee (NRC, 1999b). Alternative waste form materials, such as other glasses, glass-ceramics, and polyphase ceramics, may be better suited for these types of waste.

Off-gas particulates[4] from slurry volatilization represent another potential problem with the use of borosilicate glasses because the particulate compositions will likely differ significantly from that of the original waste feed. Thus, they may have to be blended into the waste feed at relatively low concentrations because they would be highly enriched in volatile species, such as technetium, mercury, iodine, ruthenium, cesium, boron, sodium, and possibly molybdenum. In particular, the high volatility of boron from borosilicate melts is known to induce significant losses of cesium, volatized as a cesium borate (Vance et al., 1988). Although cesium volatilization does not seem to create problems for the DWPF,[5] it may become a significant factor with future cesium-rich wastes at the Hanford Site or at the SRS, particularly if an improved method for cesium removal from the salt fraction is adopted (NRC, 2000b). Another potential problem caused by off-gas emissions in noted in Sidebar 5.4.

---

[4]Precipitated effluents in the form of fine powders or nanocrystalline condensates.

[5]Measurements in the initial waste and in the glass indicated that greater than 90 percent of the cesium-137 is incorporated in the product glass (Bibler et al., 2000).

SIDEBAR 5.4 VOLATILE EMISSIONS FROM THE DWPF MELTER LINKED TO EVAPORATOR SHUTDOWN?

One example of a problem potentially exacerbated by the carryover of off-gas emissions is the severe sodium aluminosilicate fouling of the 2H evaporators at the SRS, where at least 1,100 kilograms of aluminosilicate have been deposited in the evaporator, to depths of over a meter in some locations. Evaporators are used to concentrate supernatant liquid waste, thus conserving storage space. The carryover of condensate from the DWPF evaporator melter is also recycled back to the tank farm 2H evaporators. It is possible that entrainment or volatilization of sodium, silica, and aluminum species from the DWPF melter off-gas was in part responsible for 2H evaporator fouling. In fact, the 2H evaporator has historically processed streams high in aluminum, and in the past, small quantities of aluminosilicate buildup have been observed. Since the DWPF began operating and recycling condensate to the evaporator, the concentration of silicon in the latter has increased dramatically. However, a recent study has shown that only half of the aluminosilicate in the 2H evaporator originated from the frit carryover from the DWPF melter; the remaining half originated from a different evaporator and from laboratory analysis (Jantzen and Laurinat, in press). The EMSP is collaborating with the TFA, the Savannah River Technology Center, and the Oak Ridge National Laboratory to study the formation of aluminosilicates under conditions similar to those in the 2H evaporator.

A further issue relates to possible pre-blending of wastes from various tanks before feeding them to the melter. This method has been used at the SRS to increase the glass waste loading by smoothing the concentration of critical waste components. However, the proposed strategy for Phase I at Hanford is to empty and vitrify the wastes on a tank-by-tank basis, with little or no inter-tank blending of the wastes. Furthermore, the Hanford tank-to-tank composition variability is generally much greater than that at SRS. Thus, this strategy will require separate composition windows for waste plus glass-forming materials to be developed for each tank and may ultimately increase the overall number of glass logs.

For low-sodium wastes, the maximum waste loading is usually dictated by the concentration(s) in the waste feed of species with limited solubility in borosilicate glass, such as halides, sulfates, phosphates, chromium, and bismuth. Glass-in-glass phase separation and/or crystallization of possibly undesirable phases will occur if the solubility limits of these species are exceeded, producing a waste form that may not meet current acceptance criteria.

A good example of this limitation is the sulfate content in the waste feeds, which is determined by the various pretreatment stages (see Sidebar 5.5). The solubility of sulfate in borosilicate glass is low, so that a high sulfate content in either the HLW or the LLW streams will lead to separation of a molten sulfur-rich phase within the melter causing foaming problems. Many waste ions, including cesium, are known to partition preferentially into the sulfate phase in coexisting sulfate-silicate melts.[6] Furthermore, if the solubility limit of sulfate in

[6]P.J. Hayward, unpublished work, 2000.

SIDEBAR 5.5 IMMOBILIZATION OF HANFORD LLW

Consideration of LLW immobilization is not part of the task of this committee. Nevertheless, the quantity and composition of LLW will be dictated by pretreatment of the HLW feeds. In 1994, an amendment of the TPA between the State of Washington, the EPA, and DOE established that the LLW at Hanford will be immobilized by vitrification in borosilicate glass, using essentially the same technology proposed for HLW immobilization (TPA, 1998). The Hanford LLW stream will consist predominantly of soluble salts (nitrates, sulfates, phosphates, and carbonates) of sodium, aluminum, and potassium, together with traces of fission products and transuranic elements. The limiting factor in determining waste loading in the LLW glass will likely be the sulfate content. This limit will produce an estimated 20 percent increase in the required glass volume for LLW immobilization, compared to the volume that would be required if the LLW sodium content were the limiting factor (Pegg, 2000).

HLW or LLW glass is exceeded inadvertently, separation and accumulation of a molten sulfate or sulfide phase at the melt surface could also cause enhanced corrosion of the upper electrode(s) and refractories. The alloy used for electrodes, Inconel-690®, is known to be susceptible to attack from sulfur compounds, particularly under reducing conditions.

Long-Term Research Needs

Basic long-term research in material sciences is needed to seek alternative waste form materials, such as glass-ceramics and polyphase crystalline ceramics, for producing acceptable immobilized waste with higher concentrations of HLW from variable-composition feeds. Descriptions of many previously developed alternative waste form materials for HLW can be found in the relevant literature (Hayward, 1988; Lutze and Ewing, 1988; Donald et al., 1997). Further development of some of these materials could allow this goal to be achieved.

Further research is needed to identify more economic alternatives to borosilicate glass for immobilizing LLW in Hanford. One example could be to investigate alternative waste form materials (e.g., cement-based materials) that can incorporate higher sulfate concentrations than are possible with borosilicate glass.[7]

In the event of future immobilization of Hanford LLW in Joule-melted borosilicate glass, research will be needed to study the corrosion mechanism(s) of Inconel-690® in glass melts containing high sulfur concentrations under various redox conditions, with the goal of minimizing corrosion and/or identifying more corrosion-resistant alloys.

Crystal Content of the Borosilicate Glass Matrix

The current stipulations for the SRS borosilicate glass waste form state that the waste glass should not contain any significant degree of crystallinity or glass-in-glass phase separation (Jantzen et al., 1999). Crystallization can occur during cooling of molten HLW glass if there is sufficient overlap between the temperature ranges for substantial crystal nucleation and crystal growth. Potential crystalline phases appearing in SRS and Hanford wastes glasses are spinels ($A^{2+}[B^{3+}]_2O_4$, e.g., $NiFe_2O_4$), clinopyroxenes (principally acmite $NaFeSi_2O_6$), alkali aluminosilicates (principally nepheline $NaAlSiO_4$, albite $NaAlSi_3O_8$,

---

[7]The committee is aware of the difficulties of reversing the TPA decision to vitrify LLW (TPA, 1998). However, the EMSP should investigate alternative waste forms as part of a contingency approach to the current baseline program.

which can occur in partial solid solution with nepheline, and eucryptite $LiAlSi_3O_8$), lithium silicates (e.g., Li_2SiO_3), cristobalite (SiO_2), hematite (Fe_2O_3), and zircon ($ZrSiO_4$). Some projected precipitations in INEEL glass compositions are nepheline, fluorapatite ($Ca_{10}[PO_4]_6F_2$), lithium phosphate ($LiPO_4$), baddeleyite (ZrO_2) and alkali aluminosilicate sulfides. Above the glass transition temperature of HLW glasses, canister centerline cooling rates of approximately 0.05°C per second permit rapid-nucleating and rapid-growing phases, such as spinels and nepheline, to precipitate but disfavor difficult-to-nucleate or slow-growing phases like acmite or zircon. Spinels have been identified as major crystalline phases in glass after heat treatment at temperatures between 500°C and 900°C in WVDP glass compositions (Jain et al., 1993) and, as indicated earlier, may accumulate in Joule melters and be entrained in glass carried over from the melter.

Uncontrolled crystallization or phase separation of certain of these crystalline phases within the glass log during cooling has the potential for reducing the durability of the final waste form. The glass durability is determined by employing the product consistency test (PCT), which measures the release rates for sodium, lithium, silicon, and boron during water leaching. The deleterious influence on glass durability of certain crystalline phases derives from both chemical and mechanical effects on the surrounding glass: the residual glass composition is altered, and the glass matrix is stressed by the volume mismatch of the crystal and the glass space it replaces. Survey studies (Bailey and Hrma, 1995) have suggested that the residual glass composition is the major factor that controls the PCT response of glasses with durable crystalline phases.

Spinels are generally conceded (Jantzen and Bickford, 1985) to have little effect on glass durability, because they do not substantially alter the chemistry of the remaining glass-forming and modifying elements (silicon, aluminum, boron, and sodium). Moreover, spinels are characterized by their cubic crystal symmetry, which leads to near-isotropic[8] interface energies and strain distributions. Therefore, it is likely that spinels lead to equiaxed crystalline precipitate with mini-

---

[8]Isotropy refers to having equal physical properties (such as refractive index, thermal expansivity, elastic constants) in different crystallographic directions. Cubic crystals are usually more isotropic than other crystal systems with lower symmetry (tetragonal, hexagonal, orthorhombic, monoclinic, or triclinic). However, in some cases, cubic crystals can exhibit properties that are far from isotropic.

mal tetragonal distortion of the surrounding glass matrix. The reported effects of acmite and related clinopyroxenes, which are non-cubic and non-isotropic and comprise one of the most likely crystalline fractions, are controversial (Jantzen and Bickford, 1985), but the most recent Pacific Northwest National Laboratories (PNNL) studies conclude that acmite has virtually no impact on PCT release rates (Riley et al., 2001). Nepheline precipitation appears to have the most detrimental impact on dissolution of Hanford and Savannah River HLW glasses and can decrease chemical durability by several orders of magnitude (Riley et al., 2001). Precipitation of cristobalite (which is nonetheless cubic) and baddeleyite (which is nearly so) also impact durability negatively. Substantial crystallization of a number of other phases has been shown to have little or even a positive impact on glass durability.

Titanium (derived from monosodium titanate or crystalline silicotitanate ion exchangers, see Chapter 4), zirconium, phosphorous, and fluorine can all function as effective nucleating agents for crystallization and are commonly used for nucleating commercial glass-ceramics. Thus, it may be necessary to control the concentrations of these potential nucleating agents in the HLW feed to the melter to avoid crystallization of phases that could adversely affect waste form durability. The fact that some crystalline precipitations appear to have little or no adverse impact on chemical durability of HLW glasses suggests that it may be possible to relax the restriction on crystal content in the glass to accommodate a small content of crystalline phase. In turn, this could allow the waste loading in the glass log to be increased.

The committee is aware of the research efforts to this end being pursued at the SRS and at PNNL. DOE investigators are currently exploring the potential to increase waste loadings for Hanford, Savannah River, and INEEL by allowing crystalline precipitation upon cooling within the canister (Pittman et al., 2001). Other research efforts related to DOE's programs are under way at universities such as The Catholic University of America (Kot and Pegg, 2001) and the University of Missouri (Marasinghe et al., 1999, Ray et al., 1999). These research efforts could be effectively complemented by a long-term basic research program within the EMSP to obtain innovative approaches on the effect of crystal precipitation in borosilicate glass.

Long-Term Research Need

Long-term basic research is needed to broaden the envelope of acceptable borosilicate glass compositions to include a level of crystal content that does not adversely affect product durability.

Long-Term Leaching Properties of Borosilicate Glass Matrix

There has been a considerable amount of experimental research on the chemical durability (leach resistance) of various glass waste forms under a variety of hypothetical repository conditions and on glass corrosion mechanisms, alteration products, and long-term radiation-induced degradation.[9] However, as noted elsewhere (NRC, 1996a), the continued development of phenomenological models[10] that would allow glass-leaching data to be extrapolated over long time intervals would be advantageous.

Long-Term Research Needs

Long-term basic research is needed to further develop and verify phenomenological models to predict long-term leachability of borosilicate glass and other waste forms. Such a model should be developed and verified as a contingency against future waste form disposal issues, such as performance. A similar predictive model would be needed to support the possible future use of alternative waste form materials with higher waste loading, including glasses, glass-ceramics, and polyphase ceramics. The models should also include consideration of the influence on waste form durability of such factors as groundwater radiolysis and internal radiation damage (Weber et al., 1997; 1998).

Use of Unreacted Glass-Forming Chemicals Versus Premelted Glass Frit

Plans to immobilize HLW in borosilicate glass at Hanford include the option of using a melter feed containing "raw" chemicals (mainly oxides and carbonates), rather than premelted frit. Given the great variability of waste streams in Hanford, this option would avoid the tailoring of premelted frit to the different waste compositions to fall within the target composition window.

The rationale for using premelted frits in the melter feeds at DWPF was that residence time in the melter would be reduced because many of the glass-forming reactions would have been performed

---

[9]Much of this research is documented in journals, symposium proceedings, (e.g., the annual Materials Research Society "Scientific Basis for Nuclear Waste Management" proceedings) and DOE workshops.

[10]A phenomenological model is a multi-component model with predictive power that combines a series of mathematical descriptions of the individual phenomena involved. See the glossary in Appendix G.

ahead of time during frit manufacture. However, this would not be true if the overall rate of glass production were governed by the rate of waste dissolution in the melt pool or by the rate of heat transfer from the melt pool to the cold cap. There appears to be some uncertainty about this issue that should be resolved. The use of unreacted glass formers at Hanford could conceivably produce unanticipated melter problems (e.g., corrosion, foaming, and precipitation) that have not previously been encountered at the WVDP and at the SRS. For instance, the volume and complexity of the off-gas emissions could be greatly increased from volatilization of the slurry water content and from chemical breakdown of nitrates, nitrites, oxalates, and other organic molecules. These emissions can cause potential entrainment of other volatile species, including technetium, mercury, iodine, ruthenium, cesium, boron, and sodium.

Long-Term Research Need

Long-term basic research is needed to evaluate the controlling parameters of reaction rates and heat transfer processes in the melter. Results will strengthen the scientific basis for a rational choice between using unreacted glass formers and using premelted frit in waste feeds for future melter designs (including Hanford melters).

Foaming in Joule-Heated Melters

Foaming is the result of redox reactions within the melt and also the breakdown of anions, such as nitrates and carbonates, that generate gas during melting. Excessive foaming can form a physical and thermal barrier between the cold cap and the melt pool and can ultimately lead to melter shutdown. Foam formation is generally associated with highly oxidizing conditions in the melter (Jain and Pan, 2000). The foam acts as an insulating layer of bubbles between the melt pool and the newly introduced waste slurry feed, eventually forming a cold-cap "bridge" and preventing further waste feed from dissolving in the melt. Foaming also introduces the possibility of enhanced corrosion of the upper electrode(s) and refractories.

The problem of foaming in melters has been encountered at different DOE sites. At the WVDP this phenomenon has been 'accommodated' by adding a reducing agent (usually sugar) to the melter feed. The mechanism by which sugar reduces the formation of micro-bubbles during melting is not well understood. Foaming continues to be an issue at the SRS (Jain and Pan, 2000) and it is one of the anticipated problems at the Hanford Site, because of the wider waste composition range. Careful redox and rheology control is required to prevent foaming during the water boil-off stage. A further factor that would probably exacerbate any foaming tendency in Hanford is the proposed use of oxides and carbonates as glass formers, rather than frit. Any foaming tendency within the cold cap would be exacerbated by CO_2 generation from thermal decomposition of carbonates (Li_2CO_3 and Na_2CO_3) in the glass batch. Similar foaming problems were also encountered with the Joule-heated melter used for vitrifying mixed wastes at the Fernald site in Ohio. These problems were attributed to poor redox control, and also by foaming within immiscible sulfate layers that formed on the melt surface. Further details on foaming and redox control in melters are described by Jain and Pan (2000).

Long-Term Research Needs

Long-term basic research is recommended to characterize the behavior of the cold cap formed on the melt surface. Specifically, the sequence of reactions occurring in the cold cap and their influence on foaming tendency do not seem to be well characterized. The items to be evaluated include (1) the rates of water removal and breakdown of salts (e.g., nitrates, carbonates, and formates) and of organic additives (e.g., sugar, urea) used to control melt redox conditions; (2) the influence of feed chemistry, including sulfate content; and (3) possible oxygen evolution from redox reactions occurring within the melt. Thus, it may be possible to minimize or eliminate the potential for foaming using modifications to pretreatment and/or to the physical or chemical properties of the waste stream, (e.g., by pH or redox adjustment, change in the solids-liquid content, or particle size adjustment).

Precipitation of Noble Metals and Crystalline Phases in Joule-Heated Melters

Future melting campaigns at the SRS and at the Hanford Site will involve tank wastes with higher concentrations of noble metals (palladium, rhodium, and ruthenium), derived from the fission of uranium-235. Ruthenium is the most abundant noble metal in the Hanford HLW (Jain and Pan, 2000). Noble-metal precipitation within the melter could cause plating-out, short-circuiting, and downward drilling[11] of accumulated metal into the refractory floor, all of which would reduce melter life. Metallic precipitates also have the potential to cause alloying reactions with the Inconel-690® electrodes. Some or all of these problems are reported to have occurred elsewhere, such

[11]This phenomenon involves an enhanced refractory attack in a glass melter at contact sites between metallic inclusions and refractories. Typically, the attack involves gravity-assisted "drilling" of vertical holes or cavities in the refractories that constitute the melter floor.

as in the first melter used at the Pamela vitrification plant[12] (Demonie, 1996).

Many of the Hanford wastes, and future wastes to be immobilized at the SRS,[13] will also have relatively high iron, aluminum, nickel, manganese, and chromium contents, which, together with chromium oxide sludge from refractory corrosion, will likely cause precipitation of crystalline $A^{2+}B^{3+}_2O_4$-type spinel compounds. These dense insoluble phases may accumulate on the melter floor and could conceivably cause throat and/or riser blockage. Furthermore, many iron and chromium-rich spinel compounds exhibit relatively high electrical conductivities at glass-melting temperatures. Thus, their precipitation could lead to possible disruption of the electrical current distribution within the molten glass pool.

Long-Term Research Need

Modeling efforts, possibly combined with reduced-scale experiments, are recommended to study the consequences of precipitation and accumulation of noble metals and spinels on the melter floor. The ultimate goal of this long-term basic research is to increase glass production rates and prolong the operating life of the melter.

Limitations of Joule-Heated Melters in Achieving Higher Processing Temperatures

As noted previously, it may be advantageous to develop alternative glass or glass-ceramic waste form materials to the present generation of borosilicate waste glasses in order to achieve higher waste loadings, or to immobilize problematic wastes with unusual compositions. For example, this could be the case if the vitrification route is chosen for immobilizing the INEEL high-aluminum and high-zirconium calcines, tank heels, and other secondary waste streams from pretreatment and vitrification. These alternative waste form materials will likely require higher melting temperatures and, if fabricated using Joule-heated melters, may also require new electrode alloys and glass-contact refractories with improved corrosion resistance.

---

[12]The Pamela vitrification plant consists of two Joule-heated melters operated in Mol, Belgium, from 1985 to 1991. The first melter was shut down after three years as a result of electrical failure from buildup of noble metal sludge on the floor.

[13]The SRS staff has tested a wide range of glass compositions simulating the different types of waste streams projected for the next 25 years (Postles and Brown, 1991). Based on this study, no problems arising from future waste-composition variations are anticipated at the SRS.

Long-Term Research Needs

There is a need to study higher-temperature Joule-melting techniques as a step toward developing alternative glass or glass-ceramic waste forms with higher waste loadings and as a contingency against difficulties with future problematic wastes. Long-term basic research is required to identify (1) improved electrode materials, such as new alloys, ceramics, or cermets; (2) advanced refractories; and (3) alternative electrode-refractory configurations. In all cases, the primary goal is to minimize high-temperature corrosion in the presence of high concentrations of simulated waste under appropriate redox conditions.

Alternative Immobilization Processes to Joule-Heated Melting

Alternative immobilization processes may be advantageous as a contingency against unforeseen problems with continuous (Joule-melter) vitrification during the Hanford Phase I and II programs. A recent survey of waste immobilization technologies gives examples of some batch-processed alternatives to the continuous melters in current use at DOE sites, including processes such as induction melting, "cold-crucible" melting, or microwave melting (Jain, 2001). In many cases, these alternative processes can avoid some of the inherent problems with a continuous melter, including refractory corrosion and precipitation of noble metals and crystalline phases, although they would likely introduce other technical issues.

In general, the use of batch melting would allow greater flexibility in the range of compositions and temperatures for vitrification. This flexibility could be important if a glass or glass-ceramic waste form with a higher melting temperature were selected to immobilize future problematic waste streams or to increase the waste loading. Furthermore, the eventual task of decommissioning and disposal may be simpler with a smaller batch-type melter than with a continuous melter.

Some of these alternate immobilization technologies may require drying or pre-calcining of the HLW slurry before it could be blended with other glass or glass-ceramic precursor chemicals and processed. The committee notes that pre-calcination is used at the La Hague, Marcoule (Jouan et al., 1996) and Sellafield (Fairhall and Scales, 1996) vitrification facilities, albeit with an acidic waste stream (i.e., with little or no $NaNO_3$ content) and has also been performed on much of the INEEL waste.

Pre-calcination may offer a number of technical advantages, including prior removal of process off-gases, such as water and nitrogen oxides (NO_x), and possible elimination of foaming and associated problems. However, it would add a further step to the overall immobilization process and would involve handling fine powders. Other issues could include exothermic nitrate-organic reactions and formation of viscous sodium nitrate melts. While pre-calcining of the HLW feed does not, in itself, give higher waste loading, it may be a necessary step in any innovative process to achieve this goal. The latter could include batch-type processes where initial dry blending of the HLW feed with processing additives (e.g., frit or glass-forming chemicals) is required, or where it is important to minimize the initial volume of HLW feed.

Long-Term Research Needs

Long-term basic research is needed to identify and develop alternative melting techniques, including batch-type processes using concepts other than the continuous melters in current use at DOE sites for preparing waste forms with higher waste loadings. This research could include the study of drying or pre-calcining options for the waste feed.

6

Tank Closure and Other Long-Term Issues

It has long been considered impractical to dismantle and remove HLW tanks completely once they have been emptied because of the radiation exposure incurred by workers from radioactive residues and because of the overall prohibitive costs. Therefore, the tanks must be closed and left on site. To do so, the DOE must demonstrate, according to state and federal regulations (see Sidebar 6.1), that waste residues in the tanks can be declassified from HLW to "waste incidental to reprocessing." Before proceeding to tank closure, the DOE must adequately remove HLW from the tanks and from connecting pipelines, and must characterize and then properly immobilize residues within the tanks. Of the 239 tanks distributed among DOE HLW sites, only two tanks have been closed so far at the SRS. After tank closure, it is necessary to ensure that residual waste does not leave the tank boundaries and pose unacceptable environmental risks in the future. The area surrounding the tanks (the near field) must be monitored in the long term after tank closure to allow immediate remedial action in case residual waste releases threaten plants, animals, or humans. Radionuclides of primary concern for the environment, because of their long half-lives and their mobility, are technetium-99, selenium-79, iodine-129, carbon-14, and the actinides uranium, neptunium, and plutonium. Migration of contaminants into the environment can be controlled with the use of engineered barriers around the tanks.

In the committee's opinion, HLW cleanup problems have been too rigidly compartmentalized into tank problems and subsurface contamination problems, with the critical transition from the tank wall to the far field being ignored. Therefore, the committee addresses both tank closure issues and subsurface contamination problems in the near-field of the tanks, although the latter is not in the committee's statement of task.

SIDEBAR 6.1

WHAT IS ADEQUATE RETRIEVAL OF HLW?

High-level radioactive waste is defined by its origin from the reprocessing of spent reactor fuel. When HLW is retrieved from any storage system for disposal, there will be some radioactive residue in the storage system. Criteria have been developed for use in determining that these residues are not HLW, but WIR. The WIR criteria are presented in DOE G 435.1-1, Implementation Guide for use with DOE M 435.1 (DOE, 1999). Briefly stated, the WIR criteria are as follows:

1. **The waste has been processed (or will be further processed) to remove key radionuclides to the maximum extent that is technically and economically practical.**
2. **The waste is to be managed, pursuant to DOE's authority under the US Atomic Energy Act, so that safety requirements comparable to the performance objectives in 10 CFR Part 61, Subpart C, are satisfied.***
3. **The waste will be incorporated in a solid form at a concentration that does not exceed the applicable concentration limits for Class C LLW as established in 10 CFR Part 61.55.†**

The action guided by these criteria is to determine when HLW has been removed adequately so that residues may be treated as other than HLW and be disposed of on-site. Further details can be found in a previous NRC report (NRC, 2000b).

***Title 10 CFR Part 61 describes licensing requirements for classification of LLW sent to land disposal facilities. Subpart C describes the performance objectives of such waste.**

†Title 10 CFR Part 61.55 describes technical requirements for land disposal facilities. The regulation lists the upper limits of concentration for long-lived radionuclides and for other radionuclides, which constitute the limits for Class C LLW.

The issues discussed in this chapter are the following:

- removal of HLW from the tanks;
- removal of HLW from pipelines;
- characterization of residual waste;
- immobilization of residual waste within the tank;
- near-field monitoring after tank closure; and
- near-field containment barriers.

The issues of near-field monitoring and subsurface barriers are also addressed (often in greater detail) in other NRC reports (NRC, 1996c, 1997b, 2000a, 2000b, 2000d).

Tank Closure and Other Long-Term Issues

During its tour of the Hanford Site and of the SRS, and after discussion with site personnel, the committee identified research needs related to tank closure and other long-term issues. **The objective of the recommended long-term basic research for tank closure (and other long-term issues) is to provide the scientific basis to develop innovative methods to achieve tank closure and non-invasive monitoring of the near-field areas, as well as improved containment barriers.**

Removal of HLW from the Tanks

Because most of the HLW is retrieved from the tanks, the residual waste will likely consist of solids firmly attached to the tank surfaces as a "crust" or "hard-heel." These residual wastes present a more difficult challenge to retrieve than the bulk of the waste. First, hard-heel materials require very aggressive techniques for removal and transport, such as pulsed-action sprays, chemical dissolution approaches, and mechanical scraping technologies. The use of these processes could damage the tanks, allowing their contents to leak into the subsurface. Second, it is necessary to minimize water usage to limit secondary waste and tank leakages. Third, residual waste in tanks can be very difficult to access, since the interiors of some of the tanks are encumbered with pipes, in-tank equipment, or cooling coils. In some DSTs, waste has leaked into the "annulus," the confined space between the primary liners and the secondary containment pans. Access and removal of waste that has leaked into the annulus present a serious technological challenge (see Figure 6.1).

Long-Term Research Needs

Long-term basic research is needed to identify methods for the adequate removal of HLW from tank surfaces and from the annulus. New techniques should reduce the risk of secondary waste leakage and could use semiautonomous retrieval methods. The use of untethered robots is an example of a highly desirable basic research field that could lead to effective solutions where conventional engineering approaches often involve prohibitive costs (see Sidebar 6.2). Such robots could assist in retrieval operations by removing blockages, so that areas can be pumped, chipping off hard patches of HLW or rounding up pieces of equipment for possible retrieval and reuse.

FIGURE 6.1 HLW leak in the confined space between the primary liner and secondary pan, also called annulus, of Tank 16 at the SRS. The distance between the primary liner and the secondary pan in this type of tank (type II) varies between 2 and 6 feet. The secondary pan is only 5 feet high. SOURCE: DOE-Tanks Focus Area.

Removal of HLW from Pipelines

In addition to the waste in the tanks, the Hanford Site must also clean up waste from a maze of underground pipes. Some of the pipelines have also been plugged for a long time due to particle settling, phase changes, or reactions accompanied by precipitation or gel formation that occurred during transport. According to the TFA, there is also an increasing potential that the transfer lines currently in place will become plugged (DOE-TFA, 2000b). Currently, operations involving the cleanup of pipelines often involve manual intervention; hence they are extremely expensive. As recognized by the TFA, methods are needed to accurately locate pipelines and blockages and to unplug lines with devices that will not cause damage. Many of the activities related to pipeline cleanup belong to the province of applied research and engi-

SIDEBAR 6.2 EXAMPLE OF INNOVATIVE BASIC RESEARCH NEEDED TO AID IN TANK CLOSURE: UNTETHERED SEMIAUTONOMOUS METHODS

A tether from a robot to a controller can be a severe limitation in areas with restricted access. For example, the National Aeronautics and Space Administration (NASA) has had considerable experience with robots in severe environments. In 1992, NASA tried to send a tethered robot into a volcano crater in Antarctica. It experienced numerous problems with the connecting cable. Eventually, the entire mission was lost when the connecting fiber-optic cable broke (San Francisco Chronicle, 1993). The environment in an HLW tank is also severe, and access is also greatly restricted. An untethered robot, which could remove blockages in a tank or pipeline, retrieve lost equipment, and remove hot spots, could be of great value. Furthermore, an untethered robot would greatly simplify the decontamination requirements after retrieval.

Untethered semiautonomous robots could have many applications in the domain of HLW. For example, it may be possible to send such robots through piping or into processing vessels to monitor routine operations, locate problems in the walls of the pipe or vessel, and locate dangerous chemical conditions within the processing system. Another possible application for these robots is the investigation of soils surrounding the tanks. A burrowing untethered semiautonomous robot could sample soil chemistry and image the soil in front of it.

The use of untethered semiautonomous devices in tanks could lead to a more complete sampling of the tank contents and to a significant decrease of radiation exposure to workers, costs, risk of unexpected buildup of hazardous conditions, and turnaround time for analytical data.

Some desirable features of untethered semiautonomous tank investigation and sampling are the following:

1) **relatively low-data-rate, low-frequency, electromagnetic or seismic telemetry system for control of the robot and for transmitting processed data outside the tank to a distant operator;**
2) **wide variety of sensors installed for electrical, electromagnetic, acoustic imaging, and measurements of chemical and mechanical properties;**
3) **novel means of movement, including the ability to crawl along tank walls, move over the fluid surface, travel through the fluid, and possibly burrow through consolidated materials. For instance, such movement could be accomplished through innovative propulsion means on the robot or through manipulation of a surrounding magnetic field;**
4) **novel power supplies, including possibly exterior power supplied remotely to the robot via magnetic or electric fields without any wire link; and**
5) **small size and resistance to intense radiation and corrosive chemicals.**

Conventional engineering solutions for untethered semiautonomous tools involve budgets well beyond EMSP's budget. Therefore, this is an example where highly innovative basic research is needed to develop completely new and cost-effective solutions.

neering. However, basic research needs have been identified in two domains: (1) location of the pipelines, and (2) identification of blockages and removal:

1. *Location of pipelines.* The exact location of underground pipelines is often difficult to determine, particularly in the case of stacked pipelines. Currently, a severe limitation, also faced by other commercial organizations (e.g., utilities), is that there are no technologies available to accurately locate stacked pipelines. Of course, maps with the approximate location of each pipeline are available. However, at DOE sites, lines are still often located and cleared from the surrounding soil by hand digging. Automated digging is limited because pipelines cannot be located precisely enough.
2. *Identification of blockages and removal.* Removal of waste in plugged pipelines is a further challenge. First, it is necessary to locate the blockage. Then, the blockage must be removed without damaging the pipeline. Current methods to remove waste plugs involve chemical, pressure cycling, and vibration methods (DOE-TFA, 2000b). However, these methods require applying mechanical or corrosion stress to the equipment.

Long-Term Research Needs

Long-term basic research is needed in methods for remotely imaging the precise location of pipelines, particularly in soils, which significantly absorb or scatter the remote sensing fields. Because commercial companies also share interest in this problem, there is a great potential for dual-use technology. Seismic or electric techniques could be used for remotely locating pipeline blockages. Research on the use of untethered robots or other semiautonomous devices to remove blockages is also highly desirable to reduce workers' exposure.

Characterization of Residual Waste

To meet closure requirements there must be a determination that residual waste is WIR (see Sidebar 6.1). Characterization of the remaining radioactivity (i.e., radionuclides and their chemical form) in the tank is necessary to make this determination. Detection of residual alpha-emitting transuranic elements with half-lives greater than five years and a quantitative estimate of residual inventory are particularly important. Table 6.1 shows the radionuclide inventories in the two tanks that have been closed at the SRS (tanks 17 and 20). The main radionuclides remaining in the tanks are strontium-90, cesium-137, technetium-99, and cobalt-60. The radioactivity levels for the main residual materials in an empty waste tank are in the order of tens to

TABLE 6.1 SRS Radionuclide Waste Inventories in Tanks 17 and 20

Radionuclide	SRS Tank 17 inventory[a] Curies per 1,000 gallons	SRS Tank 20 inventory[b] Curies per 1,000 gallons
Strontium-90	379	190
Plutonium-238	29.6	8.3
Cesium-137	26.2	40.9
Americium-241	17.9	1.7
Plutonium-239	6.8	3.5
Technetium-99	1.6	0.8
Plutonium-240	1.5	0.8
Cobalt-60	1.0	0.7

NOTE: Curie inventories are estimated on the basis of 1,000 gallons of residual waste. 1 curie per 1,000 gallons = 1 curie per 3,785 liters or approximately 10 million becquerels per liter.

[a]D'Entremont et al. (1997a)
[b]D'Entremont and Hester (1997b)

hundreds of curies per 1,000 gallons (10^8 to 10^9 becquerels per liter) of residual materials. The major challenge is that characterization methods must operate in high gamma fields. For instance, in the case of the WVDP tank cleanup operation, the gamma field is about 10 rad per hour on tank walls and infrastructures and about 800 rad per hour on the bottom of the tank after the water shield is removed (DOE-TFA, 2000b).

Long-Term Research Needs

Long-term basic research on methods for the characterization of tanks, crusts, or heels is necessary to determine their chemical and radionuclide (alpha, gamma, and beta emitters) contents. Research is needed on remote sensing and analytical methods to determine the chemical and radiological composition of residual waste (see also Chapter 3). Research is particularly needed for gamma-ray spectrometry and neutron counting to survey in-tank transuranic elements and gross beta counting for strontium-90 in strong gamma fields. Research on methods that do not involve in situ sampling is desirable. Consideration should also be given to research on untethered semiautonomous characterization tools.

Immobilization of Residual Waste within the Tank

Grouting is the current baseline technique to fill the volume before closure. It provides isolation against water ingression from the surface, and immobilization of residual waste. Grout consists of inert materials such as sand and gravel, with a cement-based binder. In the two closed tanks at the SRS, the three-layered backfill system used consists of a chemically reducing grout at the bottom of the tank, a controlled low-strength material occupying most of the empty space, and a high-strength grout at the top of the tank. The primary function of the chemically reducing grout is to reduce the mobility of technetium-99, the main environmental risk contributor at the SRS. Technetium-99 is much less mobile under reducing conditions.

Other near-field containment techniques for immobilizing radionuclides include in situ vitrification, solidification and stabilization, in situ redox manipulation, and bioremediation. These techniques are described in detail in a previous NRC report (NRC, 1999d).

Long-Term Research Need

After some discussion, the committee decided that this is not one of the most fertile topics for basic research because applied research and engineering activities undertaken within the TFA and the Subsurface Contaminants Focus Area are already addressing this technological need.

Near-Field Monitoring

In this report, the near field includes the tank boundaries and extends up to 100 meters into the subsurface. Monitoring techniques for the near-field and the far-field subsurface are similar. However, special requirements exist for the near-field monitoring of HLW tanks. For example, the area surrounding the tanks is a very disturbed environment, because pumps and other rotating machinery produce significant seismic noise. Also, the presence of metal on the earth's surface leads to coupling interference with electrical geophysics survey methods.

Near-field monitoring of the subsurface is important to identify movements of fluids that might indicate tank leakage. This is key during waste processing operations, as described in Chapter 3. However, after tank closure, long-term monitoring of the near-field is still important to verify that residual waste does not traverse tank boundaries and reach the subsurface. Therefore, tank boundaries must be monitored to detect corrosion, cracks, and other signs of potential loss of waste containment. When tanks have leaked, near-field monitoring is particularly difficult because of preexisting contamination in the subsurface.

Drilling technologies are currently employed for near-field subsurface monitoring. However, drilling wells in potentially contaminated sites is extraordinarily expensive. Current costs for drilling a monitoring well to a depth of 75 meters in low-to-moderately contaminated soils are typically $210,000 per well.[1] The cost for developing and drilling a specialized sampling well in a highly contaminated soil, such as a 55-meter slant well (30 degrees) under the leaking tank SX-108 in Hanford, is about $2.5 million, which does not include sample analysis.[2] Moreover, drilling wells in a potentially contaminated region may create new pathways for the flow of contaminants and water intrusion, and generate large quantities of contaminated soil that must be treated.

A further complication in the use of drilling technologies is the heterogeneity of the earth's subsurface, as shown in Figure 6.2. To sample this heterogeneity adequately, numerous wells would have to be drilled close together, with the disadvantages cited above. Push technologies have recently become a popular alternative to core drilling. They consist of pushing a rod, equipped with various sensors, into the earth. The advantages of push technologies are that costs and chances of spreading the contamination are lower than with core drilling. However, push technologies do not penetrate to sufficient depth in soil containing large cobbles or rock formations. Furthermore, they supply only point measurements and do not provide the detail shown in Figure 6.2, obtained with alternative geophysical methods.

Imaging techniques are also becoming increasingly popular to study movements of fluids in the subsurface. Currently, gamma surveys in dry boreholes as well as seismic and electrical borehole-to-borehole methods are used to produce reasonably detailed images of the ground between two wells. However, non-invasive high-resolution methods for surveying the earth from the surface need further development. A discussion of tomography approaches and other promising subsurface imaging techniques can be found in previous NRC reports (NRC, 1996d, 2000a, 2000e).

Long-Term Research Needs

Long-term basic research is needed to develop non-invasive methods to characterize the geohydrology and also the nature, movement, and projected fate of contaminants near the tanks. Research in imaging techniques is desirable in order to develop methods that penetrate soil in depth and can identify contaminants. For example, prompt gamma neutron activation is a promising technique but has a penetration range of only one meter or less, depending on the sensitivity required. Imaging tools can also be used to monitor the state of the tank bound-

---

[1]Personal communication between committee member B.K. Sternberg and G. Mitchem, Bechtel, Hanford, Washington.

[2]Personal communication between committee member B.K. Sternberg and T. Knepp, CH2M-HILL, Hanford, Washington.

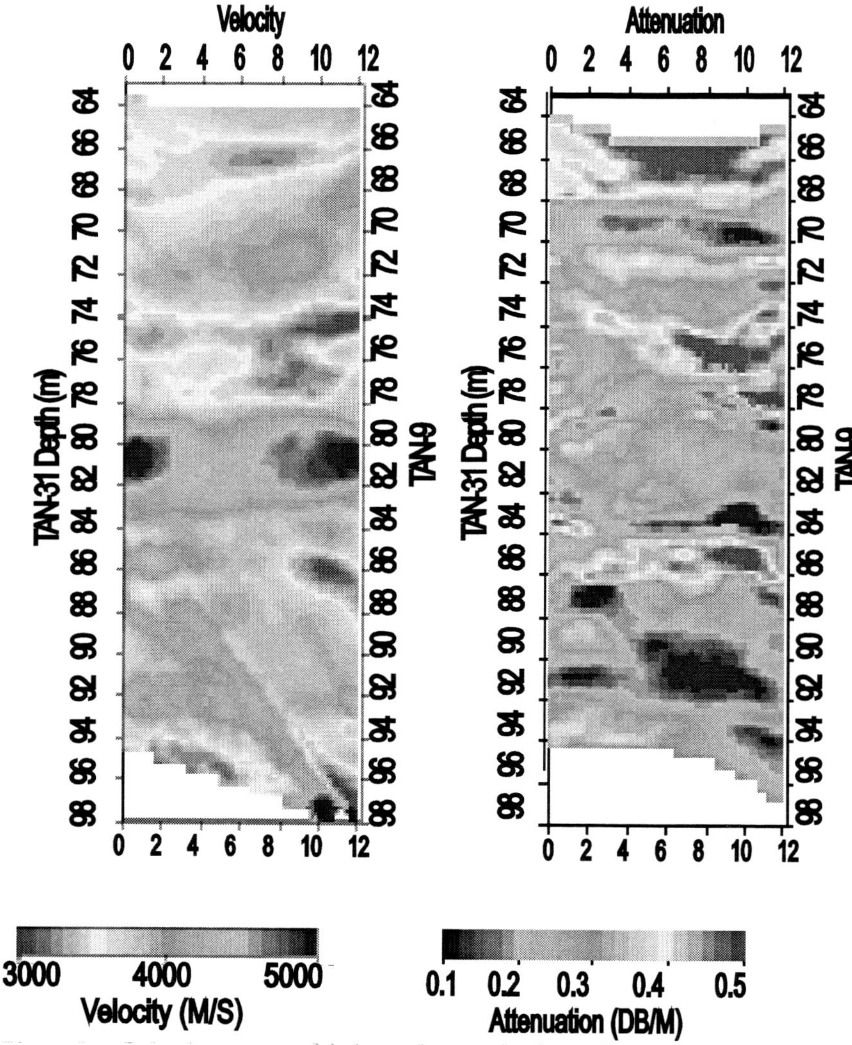

FIGURE 6.2 Representative cross-section of the earth between two wells (TAN-31 and TAN-8) at the INEEL site, showing typical heterogeneity. The cross-section image was constructed by transmitting a seismic (acoustic) signal from a fixed depth in the left well and receiving the seismic signal at many depths in the right well. The transmitter was then lowered to a new depth, and the signal was again received at many depths in the other well. The process is repeated for many transmitter locations until a large number of ray paths are measured through the earth. Data are analyzed by calculating the velocity and attenuation of the seismic signal through each cell in a grid of cells between the two wells. The resulting picture of the lateral and depth variations in velocity and attenuation can be interpreted as variations in soil (or rock) types, fracturing, and fluid movement. This cross-section image shows how complex typical soil variations are. In this black and white reproduction of the original color figure, the velocity and attenuation scales do not fully show the complete range of values (for example, black may indicate either high or low velocity/attenuation, while white represents the intermediate values). However, the purpose of this figure is only to demonstrate soil heterogeneity and the black and white scales are adequate for this purpose. The original color figure can be found in the source. SOURCE: Majer et al. (2000).

aries to identify possible damage. Investigators should consider imaging techniques used in other fields, such as medical imaging. Long-term research is also needed in modeling the transport of contaminants from the tank into the environment. Models should include the most accurate subsurface characterization parameters and estimation of the physicochemical form of the mobile contaminant. Therefore, modeling activities should be coordinated with the latest advances in subsurface contamination research and geochemical speciation studies. Because of the shared interest for remote monitoring in other fields, such as in oil and gas exploration or groundwater monitoring, there is also a great potential for science and technology transfers.

Near-Field Containment Barriers

Engineered barriers are used to reduce the flow of fluids from a contaminated area toward water supplies.[3] Engineered barriers can be physical barriers or reactive barriers. The first type consists of an impermeable wall surrounding the contaminated area; the second type consists of a permeable wall made of reactive material that allows groundwater to flow through it while it immobilizes metal or radionuclide contaminants by sorption or precipitation. Engineered barriers can be made of injected grout, various resins, or clay liners, and can be coupled with chemical treatments, freezing, or pumping techniques. Subsurface engineered barriers are under consideration at Hanford to reduce the impact to the environment of leaked waste by containing subsurface contamination within the near-field. The major drawback of current barrier technologies is that engineered barriers tend to leak even after short times. However, research is in progress in this field. For instance, a promising type of engineered barrier is a reactive barrier containing a biological layer in which microorganisms are used to immobilize contaminated material.

Long-Term Research Needs

Because barriers have a tendency to leak, long-term basic research is needed to identify methods for remotely monitoring contaminant transport through barriers. Some examples of promising remote monitoring methods are electrical and seismic geophysics methods. New materials with greatly reduced permeability or biomaterials using

---

[3]Additional detail on barrier technologies can be found in three previous NRC reports: Barrier Technologies for Environmental Management (NRC, 1997b), Groundwater and Soil Cleanup: Improving Management of Persistent Contaminants (NRC, 1999d), and Long-Term Institutional Management of U.S. Department of Energy Legacy Waste Sites (NRC, 2000d).

microorganisms to immobilize hazardous chemicals and radionuclides also offer research opportunities for engineered barriers. Moreover, future engineered barriers should protect all potential pathways, in particular underneath the tank and not just the surface surrounding it. Basic research on innovative horizontal drilling methods that do not create new pathways for flow of contaminants and new injection material to create an impermeable barrier beneath the tank is also desirable.

7

EMSP Research Portfolio: Technological Risk and Desired Attributes

This chapter provides programmatic recommendations for building an effective research portfolio. The recommendations address the issue of technological risk and desired attributes for the EMSP portfolio. These recommendations are based on the technical recommendations provided in Chapters 3 through 6 in this report, the committee's interim report (Appendix A), as well as the reports of previous committees on the EMSP (NRC, 1997a, 2000a).

Managing Technological Risk

In selecting long-term research topics for HLW management, the EMSP should take into account technological risk. A technological risk in HLW management is the risk that existing technologies will fail to accomplish goals and performance requirements set by environmental remediation policies or regulations. Neglecting to consider technological risk may lead to decreased public safety, underperformance, process failure, overruns in cost and schedule, and loss of credibility followed by a negative public perception. Examples of technological risk within the EM-HLW program are the following:

- failure to produce an acceptable glass product because of difficulties in ensuring homogeneity of the feed stock or because of deviations from the expected melter feed composition;
- premature failure of melter or of other equipment because of corrosion or mechanical problems; and

- failure of radionuclide separation processes because of unexpected secondary reactions.

One example of technological failure is the abandonment of the (large-scale) in-tank precipitation process as it was first proposed at the SRS to separate cesium from HLW (NRC, 2000b). As a consequence, DOE had to rapidly identify an alternative to this process. The EMSP, which had already invested resources for research in alternatives to the in-tank precipitation process, for instance, by funding a solvent extraction process using a calixarene to separate cesium from the waste, played an important role in the screening of processes to replace the original one.[1] Another example of an operational issue is the shutdown of the 2H evaporator and subsequent interruption of waste processing activities at the SRS because of aluminosilicate precipitation in one of the evaporator tanks (see Chapter 5, Sidebar 5.4).

The research program proposed by the committee in Chapters 3 through 6 could help the EMSP in reducing technological risk. Technological risk can be reduced by improving process effectiveness and by providing contingency approaches to baseline technologies before interferences or disruptions to current HLW management programs occur. Improving process effectiveness can lead to more robust approaches requiring fewer steps to achieve the desired result; therefore, a simple and robust process presents less technological risk compared to a multi-step process. This is because there is a finite probability of encountering a problem causing process failure in any one of the individual steps. For instance, separation techniques removing more than one targeted species in a single step pose fewer technological risks than a series of steps to achieve the same result. Of course, the single-step process might still fail, but the probability for this event is comparable to the probability of failure for each of the individual steps.

Consequently, it is important that the sites and DOE-EM collaborate to build research roadmaps identifying technological risk and including contingency approaches.[2] A fundamental part of technological risk management consists of evaluating the progress of the research and nurturing those studies yielding promising results. To this end, the committee recognizes that renewals of EMSP proposals must undergo the

---

[1]The DOE is expected to select the preferred cesium separation process during the summer of 2001. An NRC committee is providing DOE with advice for the final choice; its final report is forthcoming.

[2]This was also one of the recommendations to the EMSP by the Environmental Management Advisory Board (EMAB, 2001).

same evaluation process as new proposals so that only highly potential studies are renewed.[3]

To provide contingency approaches, it is also necessary to allocate funding for exploratory research relevant to HLW cleanup. The need for exploratory research to identify contingency approaches within the EM was also supported by previous NRC reports (NRC, 1995, 1996b, 1996c, 1997a, 1999c, 2000a). The committee recognizes that exploratory research efforts do not immediately bear fruit and that many of the alternative concepts may never be deployed in the field. However, the success of a basic research program should not be measured only by the proportion of projects that become field-applicable. Rather, it is necessary to investigate a number of possibilities before identifying a successful approach; it is accordingly impossible to predict the pathway between innovative investigations and full-scale application.

Desired Attributes of the EMSP Research Portfolio

In the selection process for proposed EMSP research topics, attention should be paid to the set of attributes listed below.

1. *Maintain long-term vision.* As noted in the interim report, the research projects that are funded should be focused not only on DOE's short-term issues but also on significant long-term problems to advance the state of knowledge well beyond the next decade. This recommendation is based on the concept of retaining a long-term vision, preserving the EMSP mission, and mitigating the technological risk of any interruption in the EM-HLW program. Such interruptions, arising from unforeseen technological problems, can be avoided by considering innovative approaches to materials, processes, and products.
2. *Maintain relevance.* Because of its mission, the EMSP should support research on the basic science underlying processes and phenomena relevant to HLW management in concert with other EM programs. EMSP basic research projects can then spark the applied research and development that will be needed for imple-

[3]In the year 2000, only 42 EMSP projects, out of the 202 coming to conclusion, have been renewed (Gilbertson, 2001).

mentation at the sites. It is thus incumbent upon both the EMSP and other programs, such as the TFA within EM, to understand the overlap of their goals and missions and yet recognize the very different nature of the projects funded. Synergies between the programs are natural and should be exploited to their fullest extent. Collaborations with foreign countries with relevant HLW research activities should be encouraged. At a minimum, the EMSP should be aware of scientific results and research trends for HLW management in countries with similar problems (see Sidebar 7.1).

3. *Provide for contingencies.* The EMSP should promote underlying science and technology that will support contingency approaches to address unanticipated difficulties encountered in baseline processes. Some fraction of EMSP projects should support exploratory and innovative research, involving non-conventional technologies, possibly leveraged from other disciplines. This recommendation was also endorsed by three previous NRC reports (NRC, 1995, 1997, 2000). All projects should represent a balanced range of research styles from large-scale teams to single investigators.
4. *Develop working relationships.* As noted in two previous NRC reports (NRC, 1997a, 2000a), the EMSP investigators should interact with the problem holders at the sites on a regular basis to learn about the nature of the problems to be solved. In return, problem holders might gain a better understanding of the scientific gaps underlying HLW problems. The committee acknowledges the fact that the EMSP workshops organized every two years are a good way to reach out to the scientific community. However, an opportunity to improve this approach would be for EMSP researchers (e.g., principal investigators, graduate students, postdoctoral fellows) to visit DOE sites regularly. Conversely, EMSP could identify liaisons among the problem holders at the sites to communicate with the investigators. As suggested in a previous NRC study, the liaisons "will not only have the greatest knowledge about the sites but will also be able to assist in integrating the results of EMSP research into the long-term EM effort" (NRC, 1997a).
5. *Prioritize objectives.* The committee recognizes that the EMSP cannot address all the proposed research areas equally in the next few years, nor can its portfolio acquire all the recommended program attributes immediately. Therefore, the EMSP should prioritize research keeping in sight the two overall motivations, a) providing contingency approaches and b) improving process

SIDEBAR 7.1 RESEARCH ACTIVITIES RELEVANT TO HLW MANAGEMENT IN FOREIGN COUNTRIES

Many countries with a nuclear defense program have produced HLW from the reprocessing of spent nuclear fuel to recover plutonium. However, the details related to these activities are highly classified. According to the information available, the former Soviet Union and the United States have HLW with similar characteristics (alkaline HLW); therefore, research and development activities in the former Soviet Union are relevant to DOE's HLW management programs. In particular, Russia is advanced in the domain of waste retrieval and chemical cleaning of the tanks. DOE's TFA is already collaborating with Russia on several topics. The Universal Solvent Extraction, or UNEX, process, an acid-side treatment for calcine developed in Russia, is currently under study at INEEL. Russia is also promoting research on mixer pump technologies for tank waste retrieval. A pulsating mixer jet pump (Flygt pump) capable of mixing waste at tank-floor level has already been deployed at the Oak Ridge National Laboratory. Other smaller collaborations between TFA and foreign countries are taking place on the following issues: resin vitrification problems (Argentina), cold crucible melter testing and pipeline unplugging (France and Russia), fluidic samplers, LLW grout formulations, saltcake dissolution and solid formation mechanisms in tanks, and residual waste sampling (United Kingdom).

Other topics relevant to HLW cleanup in which significant international experience has been gathered are alternative melting technologies, such as "in-can" or "cold crucible" melting processes in France and in the United Kingdom. Alternative radionuclide separation methods are also researched intensively abroad. For instance, France has developed a method to separate cesium from HLW through continuous precipitation and filtration of cesium tetraphenylborate (Poncelet et al., 2001).

Most of research and development efforts on HLW in foreign countries is related to the reprocessing of commercial spent nuclear fuel, which generates HLW with characteristics different than defense HLW (commercial HLW is highly acidic and relatively salt-free). However, reprocessing of commercial spent nuclear fuel has been adopted only in a few countries (France, United Kingdom, Germany, Belgium, the former Soviet Union, Japan, and India). A majority of the other countries has not decided between reprocessing and direct disposal. Research in these countries with a nuclear power program is oriented on surface storage of spent fuel, either for interim use or disposal. Research has also been performed on alternative immobilization matrices, including polyphase ceramics, for plutonium immobilization, to address anti-proliferation issues.

effectiveness, as follows. The EMSP should strive for a *balanced portfolio* addressing both problems that are *already identified* and *future potential problems*. The portfolio should have *a primary focus on identified problems* that must be solved either for efficient HLW processing operations or to provide for contingencies. Characterization, separation, and immobilization methods and processes are problem areas in which EMSP research could improve operations and provide immediate support in case of interferences or disruptions to current HLW processing programs.

At the same time, there should also be a consideration for *potential problems* that could arise or become exacerbated in the future. Examples of research objectives to address future potential problems are the following: new or better separation techniques to remove bulk non-hazardous material from the HLW stream, methods to achieve higher waste loading in immobilized waste forms, and improvements in tank closure and near-field monitoring issues.

8

Summary

In Chapters 3 through 6, the committee identifies a significant list of research topics, summarized in Table 8.1, where the EMSP can make a significant contribution in addressing HLW cleanup problems through its long-term basic research activities. The motivation driving the selection of research activities is to provide contingency approaches and to improve process effectiveness, as explained below:

- *Providing contingency approaches.* The purpose of the recommended research activities is to reduce technological risk by focusing on potential problems that could disrupt the current EM-HLW management program. Part of the recommended research for waste pretreatment and immobilization are aimed toward this objective.
- *Improving process effectiveness.* The purpose of the recommended research activities is to improve the effectiveness of the different HLW process areas (characterization, retrieval, pretreatment, immobilization, and tank closure). Research recommendations focus on better characterization methods for monitoring process functions, separation techniques to remove bulk non-hazardous material from the HLW streams, and methods to achieve higher waste loading in the immobilized waste forms.

Table 8.1 summarizes the recommended long-term research activities and identifies those providing contingency approaches and those improving process effectiveness. Some research activities could lead to both contingency approaches and process improvements. For instance, investigating the effect of crystal content in borosilicate glass is a contingency approach against problems with future waste streams but also could improve the effectiveness of the current immobilization process by increasing the waste loading. Table 8.1 also indicates whether the

TABLE 8.1 Summary of the long-term basic research objectives where the EMSP can make significant contributions in addressing HLW cleanup problems.[a]

Process Area	Long-Term Basic Research Objectives (where the EMSP can make significant contributions in addressing HLW cleanup problems)	Time Scale: Identified Problem (affecting cleanup operations now)	Time Scale: Future Potential Problem (the next few decades)	Purpose of Research Objectives: Contingency Approach (to reduce technology risks)	Purpose of Research Objectives: Improved Process Effectiveness (to reduce risks, time, and costs)
Characterization (Chapter 3)	Develop remote-sensing instruments	X			X
	Develop on-line or in situ instruments	X			X
	Improve solids-liquid separation methods	X		X	X
Retrieval and pretreatment (Chapter 4)	Improve underlying science and technology for sludge leaching operations	X		X	X
	Increase the efficiency of liquid decontamination methods	X		X	X
Immobilization (Chapter 5)	Identify alternative immobilization media to overcome limitations of borosilicate-glasses		X	X	
	Investigate the effect of increased crystal content on the durability of the borosilicate glass matrix		X	X	X
	Improve phenomenological models to predict long-term leachability of various glass waste forms		X	X	
	Evaluate advantages of using unreacted glass-forming chemicals versus premelted glass frit		X	X	X
	Understand the mechanism of foaming in Joule-heated melter		X	X	
	Mitigate effects of precipitation of noble metals and crystalline phases in Joule melters		X		X
	Improve Joule-heated melters to achieve higher processing temperatures		X	X	X
	Develop alternatives to Joule-heated melting		X	X	
Tank closure and other long-term issues (Chapter 6)	Develop highly innovative and effective retrieval methods for removal of residual HLW from the tanks	X	X		X
	Develop highly innovative and effective retrieval methods for removal of HLW materials from the pipelines	X	X		X
	Improve characterization of residual waste in tanks	X	X		X
	Develop non-invasive near-field monitoring techniques		X		X
	Improve near-field containment methods through use of barriers		X		X

[a]Details can be found in the chapters indicated in parentheses.

recommended research addresses already identified or future potential problems. Some of the recommendations may address both types of problems. For instance, improving the characterization of residual waste prior to tank closure is a problem that has been already identified by DOE. However, the first priority is to remove the bulk of the waste from the tanks, and only two tanks (see Chapter 6) have been closed so far. Nevertheless, this issue will become more pressing in the future, when retrieval of the bulk of the waste will be completed and the main task will be tank closure.

Programmatic Considerations for the EMSP Research Portfolio

In selecting long-term research topics for its portfolio, the EMSP should take into account technological risk and the following desired attributes. Examples and rationales for the recommendations are provided in Chapter 7.

Managing Technological Risk

A technological risk in HLW management is the risk that existing technologies will fail to accomplish goals and performance requirements set by environmental remediation policies or regulations. The EMSP can contribute to reducing technological risk by funding a basic research program focused on alternative processes and improved approaches to current baseline plans, as recommended in the foregoing chapters. It is important that the sites and EM collaborate to build research roadmaps identifying technological risk and including contingency approaches.

Desired Attributes of the EMSP-HLW Research Portfolio

In selecting projects for future proposal cycles, the committee recommends consideration of the following attributes for the EMSP research portfolio:

- *maintaining long-term vision,*
- *maintaining relevance,*
- *providing for contingencies,*
- *developing working relationships, and*
- *prioritizing objectives.*

The committee recognizes that the EMSP cannot address all of the recommended research areas equally in the next few years, nor can its portfolio acquire all of the recommended program attributes immediately. However, the EM cleanup program is a planned multi-decade effort costing several billion dollars. Current plans to treat and dispose of HLW are fraught with technical uncertainties, and many of the planned treatment activities are first-of-a-kind efforts presenting enormous technical challenges.

The recommendations in this report address contingency approaches or improvements to the current HLW management baseline plans. The committee is aware of DOE's commitments and of the difficulties in implementing changes, both from a technology and regulatory point of view. However, given the long-term duration of this planned cleanup effort, it is expected that new technologies will emerge and that greater scientific understanding will be achieved in the next decades. The EMSP is the ideal setting in which to develop truly innovative approaches to the management of HLW in tanks and bins that could lead to scientific breakthroughs in the future. If a scientific breakthrough can help reduce risks, cleanup time, and costs, regulations can be revised accordingly and obstacles to implementing changes at the sites can be removed. In this respect, the EMSP is a small but vital element[1] to the long-term success of the EM cleanup mission.

---

[1]This conclusion was also reached by the Ad Hoc Committee on Science and Innovation of the Environmental Management Advisory Board in EMAB (2001).

References

AAAS (American Association for the Advancement of Science). 2001. AAAS Research and Development Budget and Policy Program. Definitions. [Online]. Available: http://www.aaas.org/spp/dspp/re/define.htm [April 30, 2001].

Agnew, S.F., J. Boyer, R.A. Corbin, T.B. Duran, J.R. FitzPatrick, K.A. Jurgensen, T.P. Ortiz, and B.L. Young. 1997. Hanford Tank Chemical and Radionuclide Inventories: HDW Model Rev. 4. LA-UR-96-3860. Los Alamos, N.M.: Los Alamos National Laboratory.

Bailey, A.W. and P.R. Hrma. 1995. Waste loading maximization for vitrified Hanford HLW blend. Ceramic Transactions 61:549-556.

Benedict M., T.H. Pigford, and H.W. Levi. 1981. Nuclear Chemical Engineering, 2nd ed., pp. 467. New York, N.Y.: McGraw-Hill Book Company.

Bibler, N.E., T.L. Fellinger, S.L. Marra, R.J. O'Driscoll, J.W. Ray, and W.T. Boyce. 2000. Technetium-99 and Cesium-137 Volatility from the DWPF Production Melter During Vitrification of the First Macrobatch of HLW Sludge at the Savannah River Site. In Materials Research Society Symposium Proceedings 608:697-702.

Congress (U.S. Congress). 1996. A Bill Making Appropriations for Energy and Water Development for the Fiscal Year Ending September 30, 1996, and for Other Purposes. Energy and Water Development Appropriations Act. Public Law (P.L.) 104-47, November 13, 1995.

Congress. 1982. Nuclear Waste Policy Act of 1982. This act consists of the act of January 7, 1983 (P.L. 97-425; 96 Stat. 2201), as amended by P.L. 100-203 (December 22, 1987) and P.L. 102-486 (Energy Policy Act of 1992, October 24, 1992). The act is generally classified to 42 U.S.C. 10101 and following.

D'Entremont, P.D., J.R. Hester, and T.B. Caldwell. 1997a. Characterization of Tank 17 Residual Waste. WSRC-TR-97-0066. Revision 1. Aiken, S.C.: Westinghouse Savannah River Site Company.

D'Entremont, P.D. and J.R. Hester. 1997b. Characterization of Tank 20 Residual Waste. WSRC-TR-97-0267. Revision 0. Aiken, S.C.: Westinghouse Savannah River Site Company.

Demonie, M. 1996. Experience in Belgium. In Glass as a Wasteform and Vitrification Technology, pp. E40-41. Washington, D.C.: National Academy Press.

DOE (U.S. Department of Energy). 1997. Record of Decision for the Tank Waste Remediation System, Hanford Site, Richland, Washington. ROD 6450-01-P. February 20. Washington, D.C.: U.S. Department of Energy.

DOE. 1999. Implementation Guide for Use with DOE M 435.1. July 9, 1999. Washington, D.C.: U.S. Department of Energy.

DOE. 2000. DOE Research and Development Portfolio: Environmental Quality. Volume 2 of 4. February. Washington, D.C.: U.S. Department of Energy.

DOE-EMSP (Environmental Management Science Program). 2000. Reports and Publications. Environmental Management Science Program Annual Report. FY 2000. DOE/EM-0569. [Online]. Available: http://emsp.em.doe.gov/pdfs/2000annual.pdf [May 23, 2001].

DOE-EMSP, 2001. Environmental Management Science Program. Science Advancing Solutions for Environmental Cleanup. [Online]. Available: http://emsp.em.doe.gov [May 8, 2001].

DOE-ESPIP (Efficient Separations Program-Integrated Program. Table of Content. [Online]. Available: http://www.nttc.edu/env/ESPIP/ESPIP_TOC.html [May 8, 2001].

DOE-OCRWM (Office of Civilian Radioactive Waste Management). 1999. Waste Acceptance System Requirements Document. Revision 3. DOE-RW-0351. April. Washington, D.C.: U.S. Department of Energy.

DOE-ORP (Office of River Protection). 2000. DOE Reaches Cleanup Milestone in Hanford. In Hanford Press Release 2000 [Online]. Available: http://www.hanford.gov/press/2000/orp/111500.orp.html [June 5, 2001].

DOE-TFA (Tanks Focus Area). 2000a. TFA Multiyear Program Plan FY01-FY05 [Online]. Revised October 16. Available: http://www.pnl.gov/tfa/program/fy01mypp/index.htm [June 5, 2001].

DOE-TFA. 2000b. FY2001-FY2003 Multiyear Technical Responses [Online]. Available: http://www.pnl.gov/tfa/program/fy01techresp/index.stm [June 5, 2001].

DOE-TFA, 2001. Tanks Focus Area. Hanford Site. Waste Tanks. [Online]. Available: http://www.pnl.gov/tfa/sites/hanford/index.stm [June 5, 2001].

Donald, I.W., B.L. Metcalfe, and R.N.J. Taylor. 1997. Review: the Immobilization of High-Level Radioactive Wastes Using Ceramics and Glasses. Journal of Materials Science 32:5851-5887.

EMAB (U. S. Department of Energy's Environmental Management Advisory Board), 2001. The Role and Status of Basic Science in Accomplishing the Department of Energy's Environmental Management Mission. Ad Hoc Committee on Science and Innovation of the Environmental Management Advisory Board. April, 2001. Washington, D.C.: Office of Environmental Management.

EPA-SITE (U.S. Environmental Protection Agency-Superfund innovative Technology Evaluation Program), 2001. [Online]. Available: http://www.epa.gov/ORD/SITE [May 7, 2001].

ERDA (U.S. Environmental Research and Development Administration). 1975. Final Environmental Statement, Waste Management Operations, Hanford Reservation: ERDA-1538. Richland, Wash.: U.S. Environmental Research and Development Administration.

Fairhall, G.A. and C.R. Scales. 1996. Vitrification experience in the U.K. In Glass as a Waste Form and Vitrification Technology, pp. E48-49. Washington, D.C.: National Academy Press.

Federal Register. 2001. Office of Science and Office of Environmental Management; Office of Science Financial Assistance Program Notice 01-16: Environmental Management Science Program: Basic Science Research Related to High Level Radioactive Waste. Tuesday, January 16. Notices. Federal Register 66(10): 3564-3571.

Gephart, R.E. and R.E. Lundgren. 1998. Hanford Tank Cleanup: A Guide to Understanding the Technical Issues. PNL-10773. Richland, Wash.: Pacific Northwest National Laboratory.

Gilbertson, M. 2001. Personal Communication, May 6, 2001.

Hayward, P.J. 1988. The Use of Glass Ceramics for Immobilizing High-Level Wastes from Nuclear Fuel Recycling. Glass Technology 29:122-136.

Hrma, P.R., J.D. Vienna, M. Mika, and J.V. Crum. 1998. Liquidus Temperature Data for DWPF Glass. PNNL-11790. Richland, Washington: Pacific Northwest National Laboratory.

Huntoon, C.L. 1999. Letter to Bruce Alberts, president of the National Academies, requesting the present study. August 12, 1999. Washington, D.C.: U.S. Department of Energy.

Huntoon, C.L. 2000. Approval of the Preferred Alternative for the Idaho High-Level Waste and Facilities Disposition Environmental Impact Statement. Memorandum from C.L. Huntoon to B. Cook, Manager of the Idaho Operations Office, October 20, 2000. Washington, D.C.: U.S. Department of Energy.

IEER (Institute for Energy and Environmental Research). 1996. Science for Democratic Action Vol. 5 No.1. The Purex Process. [Online]. Available: http://www.ieer.org/sdafiles/vol_5/5-1/purexch.html [April 6, 2001].

Jain, V., S,M. Barnes, I. Joseph, L.D. Pye, I.L. Pegg, and P.B. Macedo. 1993. Impact of heat-treatment on the chemical stability of nuclear waste glasses produced at the West Valley Demonstration Project (WVDP). Ceramic Transactions 30:367-70.

Jain, V. and Y-M. Pan. 2000. Glass Melt Chemistry and Product Qualification. September 2000. CNWRA 2000-05. San Antonio, TX: Center for Nuclear Waste Regulatory Analysis.

Jain, V. 2001. Survey of Waste Solidification Process Technologies. NUREG/CR-6666, CNWRA 98-005. Washington, D.C.: U.S. Nuclear Regulatory Commission.

Jantzen, C.M. and D.F. Bickford. 1985. Leaching of Devitrified Glass Containing Simulated SRP Nuclear Waste. In Scientific Basis for Nuclear Waste Management, VIII, pp. 135-146. C.M. Jantzen, J.A. Stone and R.C. Ewing (eds.). Pittsburgh, Pa.: Materials Research Society.

Jantzen, C.M., N.E. Bibler, D.C. Beam, C.L. Crawford, and M.A. Pickett. 1993a. Development and Characterization of the Defense Waste Processing Facility (DWPF) Environmental Assessment (EA) Glass Standard Reference Material (U). A paper proposed for publication in Environmental and Waste Management Issues in the Nuclear and Ceramic Industry. WSRC-MS-93-124. Aiken, S.C.: Westinghouse Savannah River Company.

Jantzen, C.M., N.E. Bibler, D.C. Beam, C.L. Crawford, and M.A. Pickett. 1993b. Characterization of the Defense Waste Processing Facility (DWPF) Environmental Assessment (EA) Glass Standard Reference Material, Report WSRC-TR-92-346, Rev. 1, Aiken, S.C.: Westinghouse Savannah River Company.

Jantzen, C.M., K.G. Brown, T.B. Edwards, J.B. Pickett. 1998. Method of Determining Glass Durability (THERMO™). U.S. Patent 5,846,278. December 8, 1998. Washington, D.C.: U.S. Patent and Trade Office.

Jantzen, C.M., K.G. Brown, and T.B. Edwards. 1999. Predicting Phase Separation in Nuclear Waste Glasses (U). A paper for presentation at the 101th American Ceramic Society Annual Meeting in Indianapolis, Indiana, April 28, and for publication in Symposium on Waste Management Technologies in Ceramic and Nuclear Industries.WSRC-MS-99-00333, Revision 1. Aiken, S.C.: Westinghouse Savannah River Company.

Jantzen, C.M. and J.E. Laurinat. (in press). Thermodynamic Modeling of the SRS Evaporators; Part 1. The 2H and 2F Systems. WSRC-TR-2000-00293. Aiken, S.C.: Westinghouse Savannah River Company.

Jouan, A., T. Flament, and H. Binninger. 1996. Vitrification Experience in France: Development and perspectives. In Glass as a Wasteform and Vitrification Technology, pp. E42-45. Washington, D.C.: National Academy Press.

Kot, W.K. and I.L. Pegg, 2001. Glass Formulation and testing with RPP-WPP HLW Simulants. VSL-0IR 2540-2. February 2001. Washington, D.C.: The Catholic University of America.

Lutze, W. and R.C. Ewing. 1988. Radioactive Waste Forms for the Future. Amsterdam: North-Holland Press.

Mahoney, L.A., Z.I. Antoniak, W.B. Barton, J.M. Conner, N.W. Kirch, C.W. Stewart, and B.E. Wells. 2000. Results of Waste Transfer and Back-Dilution in Tanks 241-SY-101 and 241-SY-102. PNNL-13267. Richland, Wash.: Pacific Northwest National Laboratory.

Majer, E.L., F.J. Brockman, T.M. Daley, and J.E. Peterson. 2000. High-Resolution Definition of Subsurface Heterogeneity for Understanding the Biodynamics of Natural Field Systems: Advancing the Ability for Scaling to Field Conditions. EMSP Project 55264, Figure 5. Atlanta, Ga.: EMSP Workshop. [Online]. Available: http://emsp.em.doe.gov. [April 30, 2001].

Marasinghe, G.K., M. Karabulut, C.S. Ray, G. Allen, J.J. Bucher, N.M. Edelstein, D.K. Shuh, Y.S. Badyal, M.-L. Saboungi, M. Grimsditch, S. D. Shastri, and D. Haeffner. 1999. Effects of Nuclear Waste Components on Redox Equilibria, Structural Features and Crystallization Characteristics of Iron Phosphate Glasses. Ceramic Transactions 93:195-200.

NRC (National Research Council). 1995. Allocating Federal Funds for Science and Technology. Washington, D.C.: National Academy Press.

NRC. 1996a. Glass as a Waste Form and Vitrification Technology: Summary of an International Workshop. Washington, D.C.: National Academy Press.

NRC. 1996b. Environmental Management Technology-Development Program at the Department of Energy. Washington, D.C.: National Academy Press.

NRC. 1996c. The Hanford Tanks: Environmental Impact and Policy Choices. Washington, D.C.: National Academy Press.
NRC. 1996d. Rock Fractures and Fluid Flow: Contemporary Understanding and Applications. Washington, D.C.: National Academy Press.
NRC. 1997a. Building an Effective Environmental Management Science Program. Final Assessment, Washington, D.C.: National Academy Press.
NRC. 1997b. Barrier Technologies for Environmental Management. Washington, D.C.: National Academy Press.
NRC. 1998. Research Reactor Aluminum Spent Fuel. Treatment Options for Disposal. Washington, D.C.: National Academy Press.
NRC. 1999a. An End State Methodology for Identifying Technology Needs for Environmental Management, with an Example from the Hanford Site Tanks. Washington, D.C.: National Academy Press.
NRC. 1999b. Alternative High-Level Waste Treatments at the Idaho National Engineering and Environmental Laboratory. Washington, D.C.: National Academy Press.
NRC. 1999c. Decision Making in the U.S. Department of Energy's Environmental Management Office of Science and Technology. Washington, D.C.: National Academy Press.
NRC. 1999d. Groundwater and Soil Cleanup: Improving Management of Persistent Contaminants. Washington, D.C.: National Academy Press.
NRC. 2000a. Research Needs in Subsurface Science: U.S. Department of Energy Environmental Management Science Program. Washington, D.C.: National Academy Press.
NRC. 2000b. Alternatives for High-Level Waste Salt Processing at the Savannah River Site, Washington, D.C.: National Academy Press.
NRC. 2000c. Long-Term Research Needs for Deactivation and Decommissioning at Department of Energy Sites: Interim Report. Washington, D.C.: National Academy Press.
NRC. 2000d. Long-Term Institutional Management of U.S. Department of Energy Legacy Waste Sites. Washington, D.C.: National Academy Press.
NRC. 2000e. Seeing into the Earth. Noninvasive Characterization of the Shallow Subsurface for Environmental and Engineering Application. Washington, D.C.: National Academy Press.
NRC. 2001a. Evaluation of Criteria for Selecting a Salt Processing Alternative for High-Level Waste at the Savannah River Site. Washington, D.C.: National Academy Press.
NRC. 2001b. Research Opportunities for Deactivating and Decommissioning Department of Energy Facilities. Final Report. Washington, D.C.: National Academy Press.

Papp, I.G. 1998. Waste Feed Delivery Technical Basis, Volume II, Waste Feed Delivery Flowsheet. HNF-1939, Vol. II, Rev. 0. Richland, Wash.: U.S. Department of Energy.

Pegg, I.L. 2000. Telephone communication with selected committee members. October 24, 2000. Washington, D.C.: Vitreous State Laboratory.

Pittman, D.J., I.A. Reamer, D.K. Peeler, and T.B. Edwards. 2001. Property-Composition Relationships for the DP Glasses: Effect of Crystallization on Durability. Westinghouse Savannah River Site Company.

Poncelet, F.-J., J.-P. Moulin, M.-F Debreuille, and S. Runge. 2001. Continuous Precipitation and Filtration, a Process for Liquid Waste Management. In Proceedings of the Waste Management 2001 Conference. February 29-March 2, 2001. Tucson, Ariz.: U.S. Department of Energy.

Postles, R.L. and K.G. Brown. 1991. The DWPF Product Composition Control System (PCCS) at Savannah River: Statistical Process Control Algorithm. Ceramic Transactions 23:37-51.

Ramsey, W.G., C.M. Jantzen, and T.D. Taylor. 1994. Relationship Between Borosilicate Glass Composition, Structure, and Durability. Proceedings of the Environmental and Waste Management Issues in the Nuclear and Ceramic Industry. Ceramic Transactions 39:325-332.

Rapko, B.M. and G.J. Lumetta. 2000. Status Report on Phase Identification in Hanford Tank Sludges, PNNL-13394. Richland, Wash.: Pacific Northwest National Laboratory.

Ray, C.S., X. Fang, M. Karabulut, G.K. Marasinghe, and D.E. Day. 1999. Iron Redox Equilibria and Crystallization of Iron Phosphate Glasses. Ceramic Transactions 93:187-194.

Riley, B.J., J. Rosario, and P.R. Hrma, 2001. Impact of High-Level Waste Glass Crystallization on PCT Response. PNNL-13491. Richland, Wash.: Pacific Northwest National Laboratory.

San Francisco Chronicle (The). 1993. NASA Cancels Robot's Mission. Chronicle Wire Service. January 4, 1993. 01/04/93 S. F. Chron. A6. Greenbelt, Md.

Swanson, J.L. 1993. Clean Option: An Alternative Strategy for Hanford Tank Waste Remediation. PNL-8388 Volumes 1 and 2. Pacific Northwest Laboratory operated by the U.S. Department of Energy by Battelle Memorial Institute. Richland, Wash.: Pacific Northwest National Laboratory.

Thompson, M.C. 1998. Pretreatment/Radionuclide Separations of CS/TC from Supernates. WSRC-MS-98-00601. Aiken, S.C.: Savannah River Technology Center.

TPA (Tri-Party Agreement). 1998. Hanford Federal Facility Agreement and Consent Order by the Washington State Department of Ecology, U.S. Environmental Protection Agency, U.S. Department of Energy as Amended Through December 31, 1998. 89-10 Rev. 5. [Online]. Available: http://www.hanford.gov/tpa/coverpg.htm [December 29, 1999].

Vance, E.R., P.J. Hayward, and R.F. Hamon. 1988. Volatile Losses from Sphene Glass-Ceramic and Borosilicate Glass Melts. Journal of the American Ceramic Society 71(7):C318-320.

Waite, J.L. 1991. Tank Wastes Discharged Directly to the Soil at the Hanford Site: WHC-MR-0227. Office of Environmental Restoration. Richland, Wash.: U.S. Department of Energy.

Weber, W.J., R.C. Ewing, C.A. Angell, G.W. Arnold, A.N. Cormack, J.M. Delaye, D.L. Griscom, L.W. Hobbs, A. Navrotsky, D.L. Price, A.M. Stoneham, and M.C. Weinberg. 1997. Radiation effects in glasses used for immobilization of high-level waste and plutonium disposition. Journal of Materials Research 12(8):1946-1978.

Weber, W.J., R.C. Ewing, C.R.A. Catlow, T. Diaz de la Rubia, L.W. Hobbs, C. Kinoshita, J.J. Matzke, A.T. Motta, M. Nastasi, E.K.H. Salje, E.R. Vance, and S.J. Zinkle. 1998. Radiation effects in crystalline ceramics for the immobilization of high-level nuclear waste and plutonium. Journal of Materials Research 13(6):1434-1484.

Appendixes

Appendix A

Interim Report

THE NATIONAL ACADEMIES

Advisers to the Nation on Science, Engineering, and Medicine

National Academy of Sciences
National Academy of Engineering
Institute of Medicine
National Research Council

Board on Radioactive Waste Management

November 2, 2000

Dr. Carolyn Huntoon
Assistant Secretary for Environmental Management
U.S. Department of Energy
Washington, D.C. 20585

Dear Dr. Huntoon:

At your request, the National Research Council (NRC) empanelled a Committee[1] to assist the Department in developing a long-range science plan for the management of radioactive high-level waste[2] (HLW) at Department of Energy (DOE) sites.[3] Currently, the Environmental Management Science Program (EMSP) provides DOE's Office of Environmental Management (EM) "with basic research addressing fundamental issues that may be critical to advancing technologies under development, but not yet implemented."[4] The NRC provides you with this interim report to help the EMSP develop a request for proposals (RFP) aimed towards HLW management that will be published in the Federal Register by the end of this year.[5] This report has been reviewed in accordance with the procedures of the NRC[6] and reflects a consensus of the Committee.

In addition to the Committee's expertise in relevant technical disciplines and knowledge of DOE problems, several other sources of information were used to develop this interim report. The Committee reviewed a previous NRC report on the EMSP,[7] as well as three other NRC reports relevant to high-level waste management at DOE sites.[8,9,10] The Committee also held three information-gathering meetings to familiarize itself with the problems at four major HLW sites:

---

[1] Committee on Long-Term Research Needs on Radioactive High-Level Waste at Department of Energy Sites. The roster for this Committee is given in Attachment A.

[2] High-level waste is defined by the Nuclear Waste Policy Act (1982) as the highly radioactive material resulting from the reprocessing of spent nuclear fuel, including liquid waste produced directly in reprocessing and any solid material derived from such liquid waste that contains fission products in sufficient concentrations. This definition does not include DOE's spent nuclear fuel.

[3] The Committee's statement of task is given in Attachment B.

[4] EMSP Program Description (2000). In Environmental Management Science Program. [Online]. Available: http://emsp.em.doe.gov/description.htm [2000, May 31].

[5] Congress has appropriated $37 million to EMSP in fiscal year 2001, including $10 million to fund new research projects. See H.R. 4635 Department of Veterans Affairs and Housing and Urban Development, and Independent Agencies Appropriations Act, 2001.

[6] The list of report reviewers is provided in Attachment C.

[7] *Building an Effective Environmental Management Science Program: Final Assessment* (National Research Council, 1997).

[8] *Research Needs in Subsurface Science: U.S. Department of Energy's Environmental Management Science Program* (National Research Council, 2000).

[9] *Alternative High-Level Waste Treatments at the Idaho National Engineering and Environmental Laboratory* (National Research Council, 2000).

[10] *An End State Methodology for Identifying Technology Needs for Environmental Management, with an Example from the Hanford Site Tanks* (National Research Council, 1999).

2101 Constitution Avenue, NW, Washington, DC 20418 *Telephone* (202) 334-3066 *Fax* (202) 334-3077 national-academies.org

- Hanford, Washington,
- Idaho National Engineering and Environment Laboratory (INEEL),
- Savannah River, South Carolina, and
- West Valley Demonstration Project, New York.

During the first meeting, which was held on March 30-31, 2000 in Washington D.C., DOE gave the Committee an overview of EMSP research activities and HLW management problems at all of its sites. The second meeting was held in Richland, Washington on June 12-14, 2000 to receive briefings from DOE-Richland Operations Office, Pacific Northwest National Laboratory, and INEEL staff. During this meeting the Committee also toured the Hanford Site. A third meeting was held in Augusta, Georgia on August 28-30, 2000 to obtain information about the Savannah River Site and HLW immobilization issues. Presentations from staff of the DOE Savannah River Office, Westinghouse Savannah River Company, and West Valley Demonstration Project were followed by a tour of the site. Thus far, the Committee has received complete cooperation from DOE and has not been constrained in its information gathering efforts.

The Committee also reviewed the portfolio of research projects supported by EMSP since its inception in 1996.[11] The information reviewed included project titles, principal investigator names and affiliations, as well as project abstracts. The purpose of this assessment was to determine the range of research topics and issues being addressed and also to begin the process of identifying potential research gaps. As of fiscal year 2000, there are 306 research projects within the EMSP portfolio. Of these, 76 are categorized by EM under the "HLW problem area."[11] The other EM problem areas are Decontamination and Decommissioning (D&D), Subsurface Contamination, Health/Ecology/Risk, Mixed Waste, Nuclear Materials, and Spent Nuclear Fuels. A complete analysis of the EMSP research portfolio by the Committee showed that there are 92 projects closely related to HLW management problems even though some of them appeared under different EM problem areas. All 92 projects were sorted into six research fields related to HLW management (listed below) or as "other EMSP grants" if they were not directly relevant to any of the research fields. Results are summarized in Appendix D. Some of the projects may appear under more than one field because they affect several phases of the HLW management process. The six research fields relevant to HLW management are the following:

- characterization,
- waste retrieval,
- waste pretreatment,
- waste immobilization,
- tank closure, and
- in situ disposal.

[11] Environmental Management Science Program (2000). Multiple Criteria Search. In EMSP Projects Database. [Online]. Available: http://emsp.em.doe.gov/portfolio/multisearch.asp [2000, July 31].

This approach follows the DOE baseline approach to HLW management—characterize, retrieve, pretreat, immobilize, and close the tanks—and it includes the option of "in situ" disposal for at least a fraction of the waste. In this context, the Committee's definition of the "characterization research field" covers the characterization of HLW in the tanks and in the near vicinity, characterization of waste during pretreatment and immobilization, and characterization of the remaining HLW left in the tanks after retrieval. Spent nuclear fuel is not considered HLW by DOE (see footnote 2); therefore it was not addressed in the report.

Based on the information gathered to date, the Committee has identified broad research fields that would benefit from a basic science plan. The Committee does not wish to circumscribe the investigators' creativity by giving a detailed list of research projects. It is the role of the EMSP investigators to implement these broad recommendations with proposals addressing specific issues within the research fields outlined in this report. These recommendations are described in the following paragraphs.

It is the judgment of the Committee that some HLW-related problems will require further research and development to minimize risk and program cost and to improve the effectiveness of cleanup. This preliminary analysis of the EMSP portfolio and the review of the information gathered thus far has led the Committee to conclude that the RFP should solicit research projects in the following four fields, in rank order:

1. **Long-term issues related to tank closure and characterization of surrounding areas.** The figure in Attachment D shows that EMSP has awarded 21 projects related to tank closure issues. Specifically, these projects are related to the decontamination of HLW tanks and the characterization of soils around the tanks. However, as noted by a previous National Research Council report,[8] there appears to be only a small number of projects addressing radionuclide and metal contamination problems in the near-field around the tanks, as well as a lack of projects concerning engineered surface or subsurface barriers. Moreover, very few projects are devoted to the characterization of the HLW remaining in the tanks after retrieval. Therefore the Committee reiterates the recommendation in the previous NRC report[8] to strengthen research on these long-term issues. A few examples of research activities to address these issues are: i) innovative subsurface characterization methods to achieve a more effective remediation strategy, for instance using advanced imaging techniques, ii) innovative technologies to remediate small "hot spots" in the ground, and iii) characterization of the HLW remaining in or near the tanks after retrieval to facilitate tank closure.

2. **High-efficiency, high-throughput separation methods that would reduce high-level waste program costs over the next few decades.** The EMSP is supporting a significant number of separation sciences projects in the pretreatment category (Attachment D). These projects are mainly focused on understanding and controlling the speciation of elements under alkaline conditions and developing different radionuclide separation methods. However, there appears to be an insufficient focus on two important goals: high-efficiency separation and the minimization of the volume of secondary waste. The Committee recommends that the EMSP encourage proposals on separation sciences that address these two goals. The projects should directly address all types of separations: solids from liquids from gases, HLW from low-level waste, and

radionuclides from organic compounds. In the Committee's opinion, such research would help to minimize costs and improve the effectiveness of the cleanup effort. One example of a project addressing separation issues could be research on processes that remove multiple radionuclides in a single step.

3. **Robust, high loading, immobilization methods and materials that could provide enhancements or alternatives to current immobilization strategies.** Immobilization strategies involve both the immobilizing matrix and the immobilization technology. DOE currently immobilizes its HLW in a borosilicate glass matrix using slurry-fed electric (Joule) melters. In the Committee's opinion, borosilicate glasses may not be the appropriate immobilization form for all DOE waste streams, in particular for INEEL's calcined HLW and for secondary waste streams, from a risk and/or a cost point of view.[12,13] The Committee encourages research on alternative immobilization matrixes, *tailored for either HLW or low-level waste*, such as cement[14] or crystalline ceramics. Moreover, the use of Joule melters for making borosilicate glasses could limit the processes available for different waste streams leading to less efficient immobilization results. Hence, the Committee also encourages research on alternative melter techniques.[15]

4. **Innovative methods to achieve real-time, and, when practical, in situ characterization data for HLW and process streams that would be useful for all phases of the waste management program.** The Committee is aware of the numerous research projects relevant to the characterization field within the EMSP portfolio (Attachment D). Over two dozen of these projects concern real-time and/or in situ characterization techniques. However, most of the projects focus on the characterization of waste in the tanks. Very few (five projects have been identified) address the problem of characterization of the waste after retrieval, for instance the characterization of process streams and melter feeds. The Committee recommends research in this area to achieve shorter turn-around times for the analytical results, which would allow better control of HLW processing. An example of such a project is research on fiber-optical interrogation to characterize process streams.

Desired attributes of the EMSP research portfolio

In selecting the projects in this proposal cycle, the Committee believes that some attention to the following programmatic recommendations is warranted:

- As noted by previous NRC reports,[7,8] the research projects that are funded should be focused on DOE's significant long-term problems to advance the state-of-

---

[12] Lutze, W., and R.C. Ewing. 1988. *Radioactive Waste Forms for the Future.* Amsterdam: North-Holland Physics Publishing.

[13] Donald, I.W., B.L. Metcalfe, and R.N.J. Taylor. 1997. The immobilization of high-level radioactive wastes using ceramics and glasses. Journal of Materials Science, vol. 32, 5851-5887.

[14] Cementation was one of the recommended alternatives to immobilize INEEL's calcined waste; see reference 9, chapter 6.

[15] The recommendation of supporting research on alternative melter techniques was also endorsed by two other NRC reports: *Glass as a Waste Form and Vitrification Technology: Summary of an International Workshop*, (National Research Council, 1996) and reference 9.

knowledge well beyond the next decade. One significant issue is the long-term performance of materials used to immobilize HLW.[16,17]

- Because of its mission, the EMSP should promote "needs driven" or "mission-directed" basic science supporting research on fundamental processes and phenomena with potential high-impact results.[18] Such projects should have a sufficient number of single or dual-investigator teams that are exploratory and innovative, and that may use non-conventional approaches possibly borrowed from other disciplines. This recommendation was also endorsed by three previous NRC reports.[19]
- The EMSP should promote underlying science and technology parallel to "baseline" or "programmatic" approaches to enable high-level waste management efforts to be flexible in dealing with any unanticipated difficulties. A recent example of the importance of these "contingency" research activities has been the necessity for rapid identification of alternatives for separating cesium from HLW at the Savannah River Site after the in-tank-precipitation process was abandoned because of technical difficulties.[20] A considerable amount of time and money might have been saved if DOE, as far back as the early 1980's, had invested resources for research and development into alternatives to tetraphenylborate (TPB) precipitation or to better understand the mechanism of TPB decomposition.
- As recommended in two previous NRC reports,[7,8] EMSP investigators should interact with problem holders at the sites to learn about the nature of the problems to be solved. In return, the problem holders might gain a better understanding of the scientific gaps underlying the problems. The Committee therefore recommends that EMSP identifies "liaisons" among the problem holders at the sites to communicate with the investigators. The liaisons "will not only have the greatest knowledge about the sites but will also be able to assist in integrating the results of EMSP into the long-term EM effort."[7]

At the half-way point in this study, the Committee is still debating a number of issues so that it can fully address the statement of task. Although consensus has not been reached on all issues, the Committee wishes to take advantage of this interim report to inform DOE about some of the topics that are being debated. For instance, the Committee recognizes the importance of R&D in contributing to a better understanding and reduction of possible risks to the site workers and the public at large. At this point, however, the Committee has not yet converged on the specific risk-related tasks that are consistent with its understanding of the basic research element in the statement of task.

---

[16] Weber, W.J., R.C. Ewing, C.A. Angell, G.W. Arnold, A.N. Cormack, J.M. Delaye, D.L. Griscom, L.W. Hobbs, A. Navrotsky, D.L. Price, A.M. Stoneham, and M.C Weinberg. 1997. Radiation effects in glasses used for immobilization of high-level waste and plutonium disposition. Journal of Materials Research 12(8):1946-1978.

[17] Weber, W.J., R.C. Ewing, C.R.A. Catlow, T. Diaz de la Rubia, L.W. Hobbs, C. Kinoshita, Hj. Matzke, A.T. Motta, M. Nastasi, E.K.H. Salje, E.R. Vance, and S.J. Zinkle. 1998. Radiation effects in crystalline ceramics for the immobilization of high-level nuclear waste and plutonium. Journal of Materials Research 13(6): 1434-1484.

[18] In the Committee's view, basic science is defined as research that creates new generic knowledge and is focused on long-term, rather than short-term, problems. See also reference 8, page 13.

[19] See reference 7, page 3; reference 8, page 117; and *Allocating Federal Funds For Science and Technology*, pages 76-79 (National Research Council, 1995).

[20] More details are available in the NRC report *Alternatives for High-Level Waste Salt Processing at the Savannah River Site* (National Research Council, 2000).

The Committee is cognizant of the fact that the RFP planned for the end of this calendar year cannot immediately fulfill all of these essential attributes, but the EMSP should continue its endeavors to develop a portfolio having the attributes discussed in this letter. The Committee will hold two more meetings to discuss further the issues identified in this report and will develop more detailed findings and recommendations. The Committee hopes to issue its final report by June 2001.

Sincerely yours,

Michael Corradini
Chair

Attachment A: Committee Roster
Attachment B: Statement of Task
Attachment C: List of Reviewers
Attachment D: Committee's Analysis of the EMSP Portfolio

Dr. Carolyn Huntoon
November 2, 2000
Attachment A

ATTACHMENT A
COMMITTEE ROSTER

COMMITTEE ON LONG-TERM RESEARCH NEEDS FOR HIGH-LEVEL WASTE AT DEPARTMENT OF ENERGY SITES

MICHAEL CORRADINI, *Chair*, University of Wisconsin, Madison
DAVID CAMPBELL, Oak Ridge National Laboratory (retired), Oak Ridge, Tennessee
MICHELINE DRAYE, Ecole Nationale Superieure de Chimie de Paris, Paris, France
CHARLES DRUMMOND, III, Ohio State University, Columbus
PETER HAYWARD, Eutechnics Consulting, Inc., Pinawa, Manitoba, Canada
LINN HOBBS, Massachusetts Institute of Technology, Cambridge
EDWARD LAHODA, Westinghouse Science and Technology Department, Pittsburgh, Pennsylvania
ROBIN ROGERS, The University of Alabama, Tuscaloosa
BEN STERNBERG, University of Arizona, Tucson
EDWIN ZEBROSKI, Elgis Consutling, Inc., Los Altos, California

Board on Radioactive Waste Management Liaison

ROBERT M. BERNERO, U.S. Nuclear Regulatory Commission (retired), Gaithersburg, Maryland

Staff

BARBARA PASTINA, Study Director
LAURA LLANOS, Senior Project Assistant
MATTHEW BAXTER-PARROT, Project Assistant

Dr. Carolyn Huntoon
November 2, 2000
Attachment B

ATTACHMENT B
STATEMENT OF TASK

COMMITTEE ON LONG-TERM RESEARCH NEEDS FOR HIGH-LEVEL WASTE AT DEPARTMENT OF ENERGY SITES

The objective of this study is to provide recommendations to the U.S. Department of Energy's Environmental Management Science Program on the development of a long-term basic research agenda to address high-level waste problems at Department of Energy sites. The study will accomplish the following:

- Identify significant high-level waste problems that cannot be addressed effectively with current technologies.
- Recommend areas of research where the Environmental Management Science Program can make significant contributions to solving these problems and adding to scientific knowledge generally.

In recommending specific areas of research, the Committee should take into account, where possible, the agendas of other high-level waste-related research programs.

The Committee may also consider and make recommendations, as appropriate, on the processes by which (1) future research needs can be identified and (2) successful research results can be applied to the Department of Energy's high-level waste problems.

ATTACHMENT C
LIST OF REVIEWERS

This letter report has been reviewed in draft form by individuals chosen for their diverse perspectives and technical expertise, in accordance with procedures approved by the National Research Council's (NRC's) Report Review Committee. The purpose of this independent review is to provide candid and critical comments that will assist the institution in making the published report as sound as possible and to ensure that the report meets institutional standards for objectivity, evidence, and responsiveness to the study charge. The review comments and draft manuscript remain confidential to protect the integrity of the deliberative process. We wish to thank the following individuals for their participation in the review of this report:

Dr. John F. Ahearne, Sigma Xi and Duke University
Mr. Philip Clark, GPU Nuclear Corporation (retired)
Dr. Rodney C. Ewing, University of Michigan
Dr. Jane C.S. Long, University of Nevada
Dr. Juan Carlos Santamarina, Georgia Institute of Technology
Dr. Alfred P. Sattelberger, Los Alamos National Laboratory
Dr. D. William Tedder, Georgia Institute of Technology

Although the reviewers listed above have provided many constructive comments and suggestions, they were not asked to endorse the conclusions or recommendations, nor did they see the final draft of the report before its release. The review of this report was overseen by George Hornberger, appointed by the Commission on Geosciences, Environment, and Resources, who was responsible for making certain that an independent examination of this report was carried out in accordance with NRC procedures and that all review comments were carefully considered. Responsibility for the final content of this report rests entirely with the authoring committee and the NRC.

ATTACHMENT D
COMMITTEE'S ANALYSIS OF THE EMSP PORTFOLIO

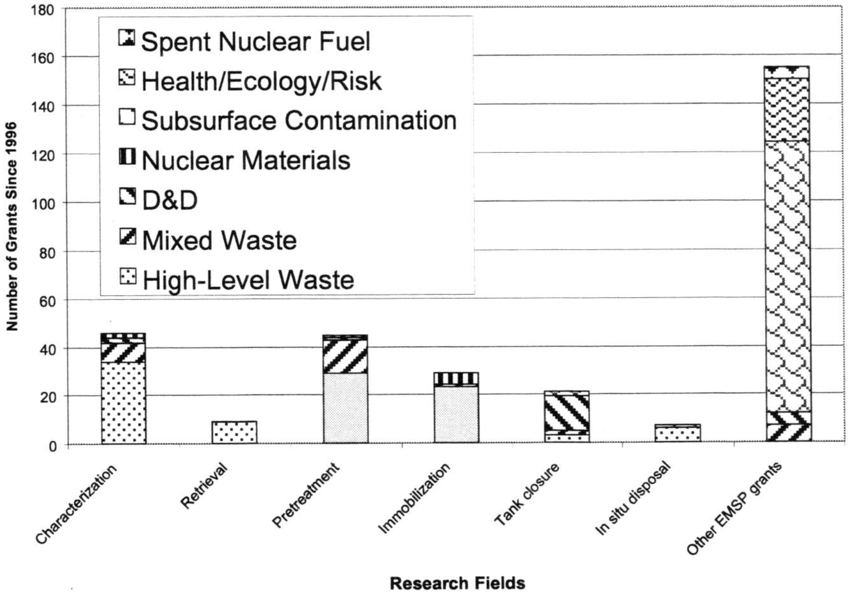

Number of projects awarded by EMSP since 1996 sorted into HLW-related research fields. The column "Other EMSP grants" includes grants that are not directly related to the management of HLW. Some projects may appear in more than one research field. The legend refers to the seven EM problem areas. Source: EMSP Online (see reference 11).

Appendix B

Analysis of the EMSP Research Portfolio

The committee generated Table B.1 during its analysis of the current EMSP research portfolio, based on the information available on the project database up to September 2000. Table B.1 does not take into account any renewal or new project introduced after that period. Only research projects directly relevant to HLW process operations are listed. The result of the analysis is shown in Appendix A, page 97. The project database is available on the World Wide Web at the following address:

http://emsp.em.doe.gov/portfolio/multisearch.asp

In the table, the title of the project is provided, with the EMSP project number shown in parenthesis. The letters in parenthesis attached to the project numbers correspond to the problem area under which the project is listed in the EMSP database. The symbols are the following:

HLW: High-Level Waste
MW: Mixed and Transuranic Waste
D&D: Deactivation and Decommissioning
NM: Nuclear Materials
SC: Subsurface Contamination
SNF: Spent Nuclear Fuel (defense-related)

As shown, many of the projects are relevant to more than one process area and have been listed more than once in the table. Therefore, the total number of projects in the table is higher than the total number of grants. The purpose of this table is to show what areas of HLW cleanup are emphasized in the EMSP research portfolio.

TABLE B.1 Analysis of the EMSP Research Portfolio

Process Area	PROJECT TITLE (EMSP Project number)
Characterization	Microstructural Properties of HLW Concentrates and Gels with Raman and IR Spectroscopies (HLW54773)
	Precipitation and Deposition of Aluminum-Containing Phases in Tank Wastes (HLW65411)
	Colloidal Agglomerates in Tank Sludge: Impact on Waste Processing (HLW54628)
	Research Program to Investigate the Fundamental Chemistry of Technetium (HLW60296)
	Imaging and Characterizing the Waste Materials Inside Tank Using Seismic Normal Modes (HLW55141)
	Acoustic Probe for Solid-Gas-Liquid Suspensions (HLW55179)
	Detection and Characterization of Chemicals Present in Tank Waste (HLW65340)
	On-line Slurry Viscosity and Concentration Measure as a Real-Time Stream Characterization Tool (HLW54890)
	Mass Spectrometric Fingerprinting of Tank Waste Using Tunable, Ultrafast Infrared Lasers (HLW 65425)
	Optically-Based Array Sensors for Selective In Situ Analysis of Tank Waste (HLW60217)
	Particle Generation by Laser Ablation in Support of Chemical Analysis of HL Mixed Waste (HLW60075)
	Correlation of Chemisorption and Electronic Effects for Metal/Oxide Interfaces (HLW65421)
	Thermospray MS Ionization Processes: Speciation, Separation, and Characterization of Organics (HLW59978)
	f-Element Ion Chelation in Highly Basic Media (HLW54595)
	Design and Development of a New Hybrid Spectroelectrochemical Sensor Project (HLW54674)
	Chemical Speciation of Inorganic Compounds Under Hydrothermal Conditions (HLW60050)
	Improved Analytical Characterization of Solid Waste Forms by Laser Ablation Technology (HLW55318)
	High Temperature Condensed Phase Mass Spectrometric Analysis (HLW60424)
	Mechanism of Pitting Corrosion Prevention by Nitrite in Carbon Steel Exposed to Dilute Salt Solutions (HLW60401)
	Mechanisms and Kinetics of Organic Aging in High-Level Nuclear Wastes (HLW65408)
	Studies Related to Chemical Mechanisms of Gas Formation in Hanford HLW (HLW54807)
	Interfacial Radiolysis Effects in Tank Waste Speciation (HLW54646)
	The NOX System in Nuclear Waste (HLW55229)
	Development of Advanced Electrochemical Emission Spectroscopy for Monitoring Corrosion in Simulated DOE Liquid Waste (HLW60219)
	Utilization of Kinetic Isotope Effects for the Concentration of Tritium (MW55103)
	Dynamic Effects of Tank Waste Aging on Radionuclide-Complexant Interactions (HLW59993)
	Analysis of Surface Leaching Processes in Vitrified HLW Using In-Situ Raman and Modeling (HLW54982)
	Development of Nuclear Analysis Capabilities for DOE Waste Management Activities (NM60077)
	Measurements and Models for Hazardous Chemical and Mixed Wastes (MW60155)
	Real-Time Broad Spectrum Characterization of Hazardous Waste by Membrane Introduction Mass Spectrometry (MW59981)
	Hazardous Gas Production by Alpha Particles in Solid Organic Transuranic Waste Matrices (MW59934)
	Development of Monitoring and Diagnostic Methods for Robots Used in Remediation of Waste Sites (D&D60040)
	High Fluence Neutron Source for Nondestructive Characterization of Nuclear Waste (HLW54751)
	Ion and Molecule Sensors Using Molecular Recognition in Luminescent, Conductive Polymers (MW55247)

Process Area	PROJECT TITLE (EMSP Project number)
	The Development of Cavity Ringdown Spectroscopy as a Sensitive Continuous Emission Monitor for Metals (MW60070)
	Microsensors for In-Site Chemical, Physical, and Radiological Characterization of Mixed Waste (MW60197)
	Novel Miniature Spectrometer for Remote Chemical Detection (MW60231)
	Miniature Nuclear Magnetic Resonance Spectrometer for In-Situ and In-Process Analysis and Monitoring (NM60247)
	Millimeter-Wave Measurements of High Level and Low Activity Glass Melts (HLW65435)
	Interfacial Radiolysis Effects in Tank Waste Speciation (HLW54646)
	Metal Ion Analysis Using Near-Infrared Dyes and the "Laboratory-on-a-Chip" (D&D64982)
	Development of Novel, Simple Multi-Analyte Sensors for Remote Environmental Analysis (HLW65001)
	The NOX System in Nuclear Waste (HLW55229)
	Fundamental Chemistry, Characterization, Separation of Tc Complexes in Hanford Waste (HLW59990)
	Characterization of Actinides in Simulated Alkaline Tank Waste Sludges and Leach Solutions (HLW65398)
	Chemical Speciation of Strontium, Americium, and Curium in High Level Waste (HLW54621)
Total	46 (HLW=34, MW=8, NM=2, D&D=2)
Retrieval	Microstructural Properties of HLW Concentrates and Gels with Raman and IR Spectroscopies (HLW54773)
	Colloidal Agglomerates in Tank Sludge: Impact on Waste Processing (HLW54628)
	Electrically Driven Technologies for Radioactive Aerosol Abatement (HLW65328)
	On-Line Slurry Viscosity and Concentration Measurement as a Real-Time Stream Characterization Tool (HLW54890)
	Numerical Modeling of Mixing of Chemically Reactive, Non-Newtonian Slurry for Tank Waste Retrieval (HLW65371)
	Acoustic Probe for Solid-Gas-Liquid Suspensions (HLW55179)
	Mixing Processes in High-Level Waste Tanks (HLW54656)
	Mechanics of Bubbles in Sludges and Slurries (HLW60451)
	Mechanics of Bubbles in Sludges and Slurries: modeling of particulate materials (HLW60451a)
	Mixing Processes in High-Level Waste Tanks (HLW54656)
Total	9 (HLW=9)
Pretreatment	Electroactive Materials for Anion Separation-Technetium from Nitrate (HLW65409)
	Reactivity of Peroxynitrite: Implications for Hanford Waste Management and Remediation (HLW59982)
	Design and Synthesis of the Next Generation of Crown Ethers for Waste Separations (HLW55087)
	Ion Recognition Approach to Volume Reduction of Alkaline Tank Waste by Separation and Recycle HLW (HLW65339)
	Speciation, Dissolution, and Redox Reactions of Cr Relevant to Pretreatment and Separation (HLW65368)
	Dynamic Effects of Tank Waste Aging on Radionuclide-Complexant Interactions (HLW59993)
	Solution Effects on Cs Complexation with Calixarene Crown Ethers from Liquids to Supercritical Fluids (HLW65351)
	Potential-Modulated Intercalation of Alkali Cations into Metal Hexacyanoferrate Coated Electrodes (HLW60123)
	Complexants for Actinide Element Coordination and Immobilization (HLW65378)

Process Area	PROJECT TITLE (EMSP Project number)
	Polyoxometalates for Radioactive Waste Treatment (HLW54716)
	Removal of Technetium, Carbon Tetrachloride, and Metals from DOE Properties (HLW60017)
	Fundamental Chemistry, Characterization, Separation of Tc Complexes in Hanford Waste (HLW59990)
	Development of Inorganic Ion Exchangers for Nuclear Waste Remediation (HLW54735)
	Ionizing Radiation Induced Catalysis on Metal Oxide Particles (HLW54996)
	Rational Design of Metal Ion Sequestering Agents (HLW60362)
	Precipitation and Deposition of Aluminum-Containing Phases in Tank Wastes (HLW65411)
	Enhanced Sludge Processing of HLW: Hydrothermal Oxidation of Cr, Tc, and Complexants by NO3- (HLW54765)
	Colloidal Agglomerates in Tank Sludge: Impact on Waste Processing (HLW54628)
	Investigation of Novel Electrode Materials for Electrochemically-Based Remediation of H and LLW (HLW55137)
	Foaming in Radioactive Waste Treatment and Immobilization Processes (HLW60143)
	Superconducting Open-Gradient Magnetic Separation for the Pretreatment of Vitrification Feeds (HLW55294)
	Electrically Driven Technologies for Radioactive Aerosol Abatement (HLW65328)
	Actinide-Specific Interfacial Chemistry of Monolayer Coated Mesoporous Ceramics (HLW65370)
	Actinide-Aluminate Speciation in Alkaline Radioactive Waste (HLW65318)
	Characterization of Actinides in Simulated Alkaline Tank Waste Sludges and Leach Solutions (HLW65398)
	Synthesis and Characterization of Templated Ion Exchange Resins for the Selective Complexation (HLW59977)
	Photocatalytic and Chemical Oxidation of Organic Compounds in Supercritical Carbon Dioxide (MW54847)
	Chemical Speciation of Inorganic Compounds Under Hydrothermal Conditions (HLW60050)
	Isolation of Metals from Liquid Wastes: Reactive Scavenging in Turbulent Thermal Reactors (MW60326)
	Acid-Base Behavior in Hydrothermal Processing of Wastes (MW54506)
	Processing of High-Level Waste: Spectroscopic Characterization of REDOX Reactions in Supercritical Water (MW54828)
	Fundamental Chemistry and Thermodynamics of Hydrothermal Oxidation Processes (MW55276)
	Chemical Speciation of Sr, Am, and Cm in High-Level Waste: Predictive Modeling of Phase Partitioning During Tank Processing (HLW54621)
	Managing Tight-Binding Receptors for New Separations Technologies (MW54791)
	Spectroscopy, Modeling, and Computation of Metal Chelate Solubility in Supercritical CO2 (MW4942)
	f-Element Ion Chelation in Highly Basic Media (HLW54595)
	Extraction and Recovery of Mercury and Lead from Aqueous Waste Streams Using Redox-Active Layered Metal Chalcogenides (MW55012)
	De Novo Design of Ligands for Metal Separation (MW55223)
	Rational Synthesis of Imprinted Organofunctional Sol-Gel Materials for Toxic Metal Separation (MW60096)
	Adsorption/Membrane Filtration as a Contaminant Concentration and Separation Process for Mixed Wastes and Tank Wastes (MW55146)
	Development of Advanced In Situ Techniques for Chemistry Monitoring and Corrosion Mitigation in SC WO Environments (MW55171)

Process Area	PROJECT TITLE (EMSP Project number)
	Architectural Design Criteria for F-Block Metal Ion Sequestering Agents (MW54679)
	Aqueous Electrochemical Mechanisms in Actinide Residue Processing (NM59967)
	Basis for the Characterization, Separation, and Disposal of Plutonium and Other Actinides in High-Level Radioactive Waste: The Effect of Temperature and Electrolyte Concentrations on Actinide Speciation (HLW65352)
	Synthesis of New Water-Soluble Metal-Binding Polymers: Combinatorial Chemistry Approach (D&D54724)
Total	44 (HLW=29, MW=13, D&D=1, NM=1)
Immobilization	Chemical Speciation of Inorganic Compounds Under Hydrothermal Conditions (HLW60050)
	Modeling of Spinel Settling in Waste Glass Melter (HLW65422)
	Radiation Effects in Nuclear Waste Materials (HLW54672)
	Radiation Effects on Materials in the Near-Field of Nuclear Waste Repository (HLW54691)
	Analysis of Surface Leaching Processes in Vitrified HLW Using In-Situ Raman and Modeling (HLW54982)
	New Silicotitanate Waste Forms: Development and Characterization (HLW60345)
	Chemical Decomposition of HLW Storage/Disposal Glasses Under Irradiation (HLW55188)
	The Influence of Radiation and Multivalent Cation Additions on Phase Separation and Crystallization of Glass (HLW59827)
	Physical, Chemical, and Structural Evolution of Zeolite Waste Forms from Metakaolinite and Calcined HLW (HLW65366)
	Investigation of Microscopic Radiation Damage in Waste Forms Using ODNMR and AEM Techniques (HLW55367)
	Stability of High-Level Waste Forms (HLW60020)
	Research Program to Investigate the Fundamental Chemistry of Technetium (HLW60296)
	High Temperature Condensed Phase Mass Spectrometric Analysis (HLW60424)
	Millimeter-Wave Measurements of High-Level and Low-Activity Glass Melts (HLW65435)
	Developing a Characterization, Separation, and Disposal of Pu and Other Actinides in HLW (HLW65352)
	Radiation Effects on Transport and Bubble Formation in Silicate Glasses (HLW60313)
	Ion-Exchange Processes and Mechanisms in Glasses (HLW60362)
	Complexants for Actinide Element Coordination and Immobilization (HLW65378)
	Chemical and Ceramic Methods Toward Safe Storage of Actinides Using Monazite (NM55094)
	An Alternative Host Matrix Based on Iron Phosphate Glasses for the Vitrification of Specialized Nuclear Waste Forms (MW55110)
	Fundamental Thermodynamics of Actinide-Bearing Mineral Waste Forms (NM60118)
	Rational Design of Metal Ion Sequestering Agents (HLW60362)
	Distribution and Solubility of Radionuclides and Neutron Absorbers in Waste Forms for Disposition of Plutonium Ash and Scraps, Excess Plutonium, and Miscellaneous Spent Nuclear Fuels (NM60387)
	Quantifying Silica Reactivity in Subsurface Environments: Controls of Reaction Affinity and Solute Matrix on Quartz and SiO2 Glass Dissolution Kinetics (HLW55042)
	Thermodynamics of the Volatilization of Actinide Metals in the High-Temperature Treatment of Radioactive Wastes (NM60319)
	Actinide-Specific Interfacial Chemistry of Monolayer Coated Mesoporous Ceramics (HLW65370)

Process Area	PROJECT TITLE (EMSP Project number)
	Physical, Chemical, and Structural Evolution of Zeolite-Containing Waste Forms Produced from Metakaolinite and Calcined High-Level Waste (HLW65366)
	Determination of Transmutation Effects in Crystalline Waste Forms (NM55382)
	Polyoxometalates for Radioactive Waste Treatment (HLW54716)
	Modeling of Spinel Settling in Waste Glass Melter (HLW65422)
Total	30 (HLW=24, NM=5, MW=1)
Tank closure	Rapid Migration of Radionuclides Leaked from High-Level Waste Tanks (HLW65410)
	Phase Chemistry of Tank Sludge Residual Components (HLW60403)
	Phase Chemistry of Tank Sludge Residual Components (HLW60403a)
	Contaminant-Organic Complexes, Their Structure and Energetics in Surface Decontamination Processes (D&D64947)
	Decontamination of Radionuclides from Concrete During and After Thermal Treatment (D&D64986)
	Three-Dimensional Position-Sensitive Germanium Detectors (D&D65015)
	Waste Volume Reduction Using Surface Characterization and Decontamination by Laser Ablation (D&D60283)
	Micelle Formation and Surface Interactions in Supercritical CO2 Fundamental Studies for the Extraction of Actinides from Contaminated Surfaces (D&D64865)
	Supercritical Carbon Dioxide-Soluble Ligands for Extracting Actinide Metal Ions from Porous Solids (D&D64965)
	Real-Time Identification and Characterization of Asbestos and Concrete Materials with Radioactive Contamination (D&D65004)
	Advanced Sensing and Control Techniques to Facilitate Semi-Autonomous Decommissioning (D&D55052)
	Development of Monitoring and Diagnostic Methods for Robots Used in Remediation of Waste Sites (D&D60040)
	Mineral Surface Processes Responsible for the Decreased Retardation for Enhanced Mobilization of 137Cs from HLW Tank Discharges (SC60355)
	Atmospheric-Pressure Plasma Cleaning of Contaminated Surfaces (D&D54914)
	In-Situ Spectro-Electrochemical Studies of Radionuclide Contaminated Surface Films on Metals and the Mechanism of Their Formation and Dissolution (D&D55380)
	Mechanisms of Radionuclide-Hydroxycarboxylic Acid Interactions for Decontamination of Metallic Surfaces (D&D64946)
	Optimization of Thermochemical, Kinetic, and Electrochemical Factors Governing Partitioning of Radionuclides During Melt Decontamination of Radioactively Contaminated Stainless Steel (D&D60363)
	Reactivity of Primary Soil Minerals and Secondary Precipitates Beneath Leaking Hanford Waste (SC70070)
	Spectroscopy, Modeling and Computation of Metal Chelate Solubility in Supercritical CO2 (MW54942)
	Photocatalytic and Chemical Oxidation of Organic Compounds in Supercritical Carbon Dioxide (MW54847)
	Improved Decontamination: Interfacial, Transport, and Chemical Properties of Aqueous Surfactant Cleaners (D&D64912)
	Long-Term Risk from Actinides in the Environment: Modes of Mobility (SC60015)
Total	22 (HLW=3 D&D=14 SC= 3 MW=2)

Appendix C

Spent Nuclear Fuel Issues

The DOE spent nuclear fuel includes both irradiated and unirradiated fuel from various reactors and experiments (including fuel from the Fast Flux Test Facility reactor), test reactor fuel, Navy fuel, and even damaged core debris from the Three Mile Island Unit 2 accident. This study does not consider research needs relating to the disposition of spent nuclear fuel, although DOE has ownership of substantial amounts with widely varying characteristics and integrity. These fuels are stored at Hanford, Savannah River Site, and INEEL.

A large fraction of DOE's fuel inventory consists of the 2,200 metric tons of zirconium-clad uranium metal fuel in the two Hanford K-Basins. Nearly half of this fuel is damaged, some to the point of fine scrap and sludge. There is an on-going program to collect most of this material in multi-canister overpacks. The material is then dried and rendered inert prior to its disposal in the geological repository. However, several tons of finely divided particulates are slated to be dissolved, added to the HLW tanks at Hanford, and co-disposed with the HLW.

Appendix D

EMSP Overview and HLW Research Program

The Department of Energy's EMSP was created in 1996 by the 104th Congress within the Energy and Water Development Appropriations Act to stimulate basic research for environmental cleanup of the nation's nuclear weapons complex. Congress directed the department to "provide sufficient attention and resources to longer-term basic science research which needs to be done to ultimately reduce cleanup costs . . . develop a program that takes advantage of laboratory and university expertise, and . . . seek new and innovative cleanup methods to replace current conventional approaches which are often costly and ineffective" (Congress, 1996).

The mission of the EMSP is the following:[1]

- develop a targeted, long-term basic research agenda to reduce cleanup costs and risks to workers and the public;
- bridge the gap between broad fundamental research and needs-driven applied technology; and
- serve as a stimulus for focusing the nation's science infrastructure on critical environmental problems.

Since its inception in 1996, the EMSP has awarded 306 research projects in seven environmental management problem areas, classified as follows:

1. High-level waste
2. Mixed and transuranic waste

[1]Environmental Management Science Program Annual Report. FY 2000. DOE/EM-0569. [Online]. Available: http://emsp.em.doe.gov/pdfs/2000annual.pdf [May 23, 2001].

3. Deactivation and decommissioning
4. Subsurface contamination
5. Spent nuclear fuel
6. Nuclear materials
7. Health, ecology, and risk

The year 2000 breakdown of EMSP funding by problem area are shown in Figure D.1. A selection of the "most successful projects" as defined by the EMSP, including eight projects that have been implemented into the applied TFA program, is shown in Sidebar D.1.

Office of Environmental Management and the EMSP

The EMSP is a collaborative partnership between the EM and the SC. The Office of Science and Technology (OST), within EM, and the Office of Basic Science, within SC, manage the EMSP, and the DOE's Idaho Operation Office administers it. The organization chart of the EMSP is shown in Figure D.2. The OST provides the EM with basic research, risk assessment, and technology development and deployment in nuclear weapons cleanup activities. The EMSP is in charge of the basic research part of the OST program.

The EMSP research has explicit links to problem holders including technical staff, managers, and stakeholder advisory groups at the sites. The EMSP collaborates particularly closely with applied research and development units within OST called "focus areas." The focus areas

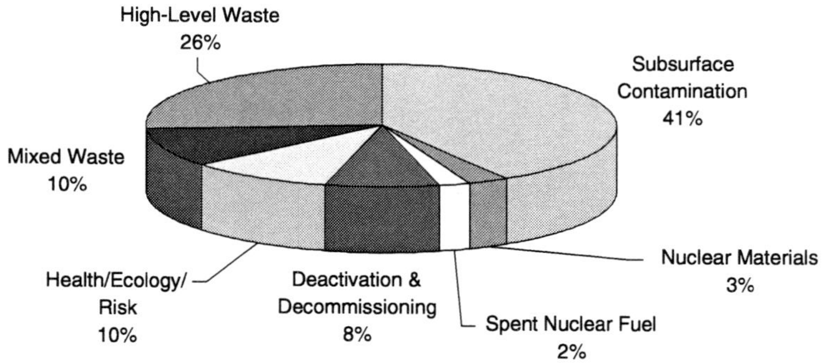

FIGURE D.1 Year 2000 breakdown of EMSP funding by problem area. The total funding through 2000 awards for the EMSP budget has been $254.3 million. In 2001 the total EMSP appropriations were $37 million. SOURCE: DOE-Environmental Management Science Program.

SIDEBAR D.1 EMSP "SUCCESS STORIES"

As identified by the EMSP, examples of successful research projects within the HLW problem area (the project identification number is shown in parenthesis) are the following:

- **Chemical Speciation of Strontium, Americium, and Curium in High Level Waste: Predictive Modeling of Phase Partitioning During Tank Processing (HLW54621)***
- **On-Line Slurry Viscosity and Concentration Measure as a Real-Time Stream Characterization Tool (HLW54890)**
- **Quantifying Silica Reactivity in Subsurface Environments: Controls of Reaction Affinity and Solute Matrix on Quartz and SiO2 Glass Dissolution Kinetics (HLW55042)***
- **Design and Synthesis of the Next Generation of Crown Ethers for Waste Separations (HLW55087)***
- **An Alternative Host Matrix Based on Iron Phosphate Glasses for the Vitrification of Specialized Nuclear Waste Forms (MW55110)**
- **Acoustic Probe for Solid-Gas-Liquid Suspensions (HLW55179)**
- **Research Program to Investigate the Fundamental Chemistry of Technetium (HLW60296)**
- **New Silicotitanate Waste Forms: Development and Characterization (HLW60345)***
- **Architectural Design Criteria for F-Block Metal Ion Sequestering Agents (MW54679)**
- **Supramolecular Chemistry of Selective Anion Recognition for Anions of Environmental Relevance (MW54864)**
- **The NOX System in Nuclear Waste (HLW55229)**
- **Reactivity of Peroxynitrite: Implications for Hanford Waste Management and Remediation (HLW59982)**
- **New Anion-Exchange Resins for Improved Separations of Nuclear Materials (MW54770)**
- **Novel Miniature Spectrometer for Remote Chemical Detection (MW60231)**
- **Foaming in Radioactive Waste Treatment and Immobilization Processes (HLW60143)***
- **Millimeter-Wave Measurements of High-Level and Low-Activity Glass Melts (HLW65435)***
- **Rational Design of Metal Ion Sequestering Agents (HLW60362)***
- **Precipitation and Deposition of Aluminum-Containing Phases in Tank Wastes (HLW65411)**

***EMSP projects now integrated within TFA programs.**

develop technologies to address radioactive and hazardous waste issues focusing on specific problems. For HLW-related research, the EMSP closely collaborates with the TFA, which is concerned with HLW in the tanks and processing operations.

Although they address the same problems, the TFA and EMSP programs are very different. The former has a short-term focus and aims at applied research. The latter addresses the more fundamental issues of underlying science and engineering technology and has a longer time frame to develop new and original science. The EMSP also col-

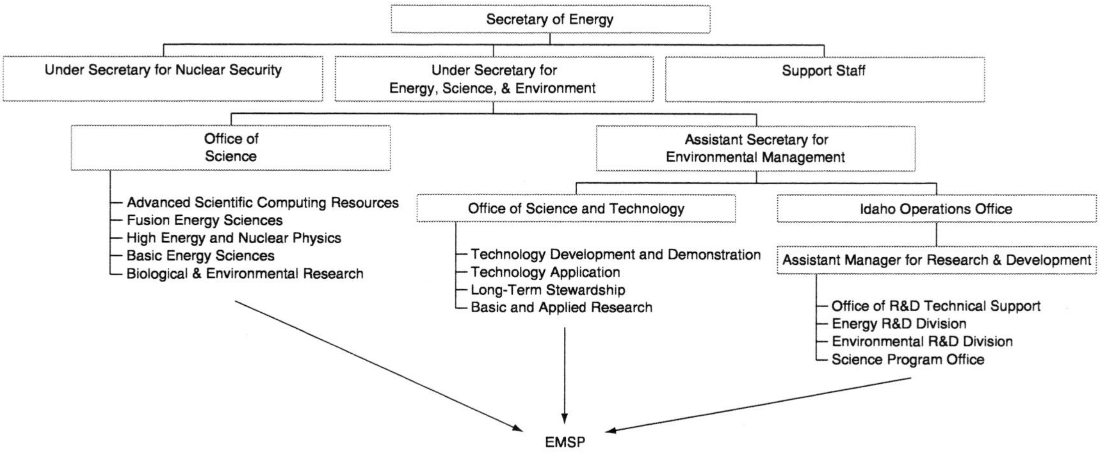

FIGURE D.2 EMSP organization chart within the Office of Environmental Management. SOURCE: DOE-Environmental Management Science Program.

laborates with the Site Technical Coordinating Groups (STCGs) in charge of compiling a list of science needs and communicating technology issues among site projects, technology providers, DOE sites, regulators, and other stakeholders.

A "science need," as defined by the EMSP, is a cleanup problem, either now or in the future, that cannot be addressed practically with current knowledge or technology. In the past, science needs were identified exclusively by stakeholder input. A stakeholder is anyone with an interest in DOE activities or anyone who may be affected by DOE activities. Previously, science needs for the EMSP[2] were drawn mainly from the STCGs and the focus areas. High-level waste research needs were usually identified by the TFA.

Figure D.3 illustrates the EMSP selection process for research awards. From the identification of needs, the EMSP develops a solicitation for research proposals among universities and national laboratories. Proposals undergo a two-step review. The first is for scientific merit by an external panel of experts under the auspices of DOE's SC. The second consists of an EM relevance review, performed in collaboration with TFA staff to ensure that the research project has a high potential to contribute, directly or indirectly, to the cleanup of the sites. The projects are funded on a three-year basis, with the possibility of renewal. Renewals are also submitted to the same two-step competitive peer review process.

---

[2]The EMSP needs database is accessible via the World Wide Web (at the URL http://emsp .em.doe.gov).

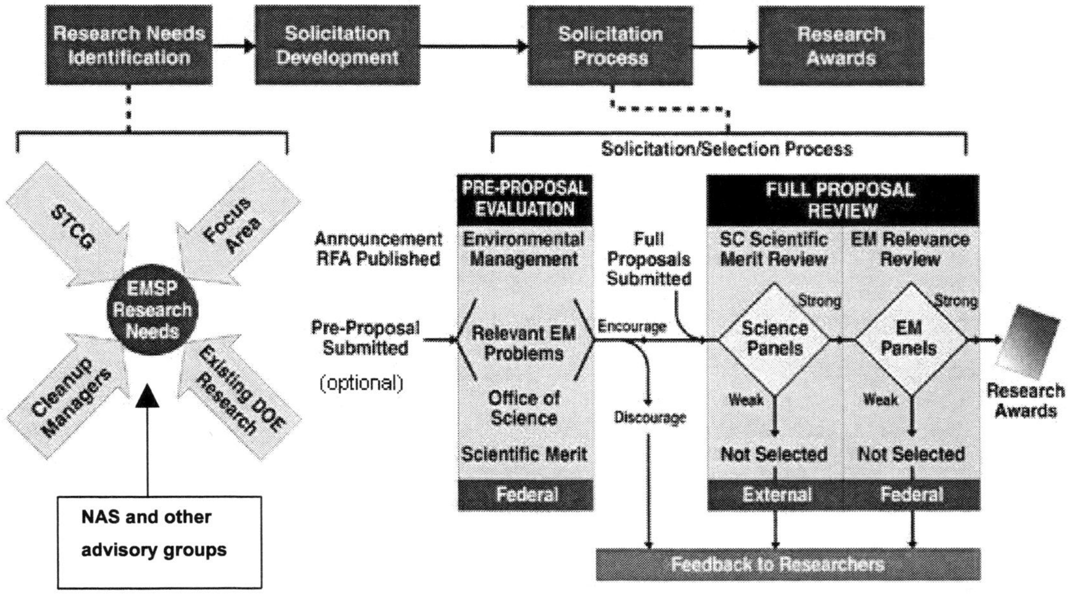

FIGURE D.3 Process for the selection of EMSP research projects. Research needs are submitted by the Site Technical Coordinating Groups (STCGs) and by the National Academy of Sciences (NAS), among others. SOURCE: DOE-Environmental Management Science Program.

Appendix E

Biographical Sketches of Committee Members

Michael Corradini, *Chair,* is a professor in the Department of Engineering Physics at the University of Wisconsin, Madison, and associate dean of the College of Engineering. Dr. Corradini's research focus is nuclear engineering and multiphase flow with specific interests that include light-water reactor safety, fusion reactor design and safety, waste management and disposal, vapor explosions research and molten core concrete interaction research, and energy policy analysis. He is a member of the American Institute of Chemical Engineers, the American Society of Engineering Education, and the American Society of Mechanical Engineers, and is a fellow of the American Nuclear Society. Dr. Corradini has received numerous awards including the National Science Foundation's Presidential Young Investigators Award, the American Nuclear Society reactor safety best paper award, and the University of Wisconsin, Madison, campus teaching award. He is the author of more than 100 technical papers and has served on various technical review committees, including the research review panel of the U.S. Nuclear Regulatory Commission and the direct heating review group. He is currently a member of the National Research Council's Electric Power/Energy Systems Engineering Peer Committee and chair of the Frontiers of Engineering Organizing Committee. Dr. Corradini was elected to the National Academy of Engineering in 1998. He received his B.S. in mechanical engineering from Marquette University and his M.S. and Ph.D. in nuclear engineering from the Massachusetts Institute of Technology.

David Campbell worked as a chemist with the Chemical Technology Division of Oak Ridge National Laboratory until his retirement in 1991 and has since consulted on nuclear fuel and radioactive waste processing. He is a member of the U.S. Department of Energy

High Level Waste Tanks Technical Advisory Panel and the subpanel on Hanford waste pretreatment. His research interests include radiochemical processing and separations chemistry. He is a member of the American Chemical Society and the American Nuclear Society, which honored him with the Special Award for Advancements in Nuclear Technology in response to Three Mile Island. Other awards include the Glenn T. Seaborg Actinide Separations Award. He served on the National Research Council's Committee on the Idaho National Engineering and Environmental Laboratory High-Level Waste Alternative Treatments. He received his B.A. in mathematics and chemistry from the University of Kansas City (Missouri) and his Ph.D. in physical chemistry from the Illinois Institute of Technology.

Micheline Draye is a professor at the École Nationale Supérieure de Chimie de Paris. She conducts research in the Laboratory of Electrochemistry and Analytical Chemistry (UMR CNRS 7575) where she is a member of the Separation Processes and Radiochemistry Group. Dr. Draye's research interests include the development of new separation processes for pollution prevention and remediation, the design of efficient extraction systems with high selectivity for hazardous ions and the investigation of the effect of radiolysis on organic ligands and ion-exchange resins. Her awards include the 1991 prize of the French Society of Nuclear Energy, a fellowship with the Commissariat à l'Énergie Atomique and a fellowship with Compagnie Générale des Matières Nucléaires, or COGEMA. Dr. Draye worked for two years as visiting scientist in the Nuclear Engineering Department of the Massachusetts Institute of Technology, after having received her Ph.D. in analytical chemistry from the University of Lyon, France.

Charles Drummond, III is a professor in the Department of Materials Science and Engineering at the Ohio State University. His research is concerned primarily with the structure and properties of amorphous solids or glasses the nature of the glassy state and the structure and crystallization of glasses and the vitrification of industrial and governmental waste to produce salable products. Dr. Drummond is a fellow of the American Ceramic Society, has been awarded its Governor's Award for Emerging Technology, and is Director of the Annual Conference on Glass Problems. He received his bachelor of ceramic engineering, bachelor of engineering physics, and degrees M.S. in Ceramic Engineering from the Ohio State University and the science master and Ph.D., in applied physics from Harvard University.

Peter Hayward is a senior consultant with Eutechnics Consulting, Inc. Dr. Hayward has more than 18 years of research experience on reactor safety issues and waste management with Atomic Energy of

Canada Limited. He is a member of the Canadian Nuclear Society and the American Ceramics Society and has authored or co-authored numerous papers on nuclear waste glasses, glass-ceramics, ceramics, and reactor safety issues. His current research interests include nuclear and toxic waste immobilization, nuclear fuel-cladding-steam interactions under reactor accident conditions, ceramic membranes, and high-temperature mineral extraction processes. Dr. Hayward received his B.Sc. in geology and chemistry from Bristol University in the United Kingdom and his Ph.D. in geochemistry and ceramic science from Birmingham University in the United Kingdom.

Linn Hobbs is professor of Materials Science at the Massachusetts Institute of Technology and was the inaugural holder of the John F. Elliott chair there. His research activities center on characterization, using electron microscopy and diffraction methods, of atomic and extended defect structures of inorganic nonmetals introduced by radiation or chemically driven compositional change and atomic-scale modeling of amorphized structures using topological approaches. He is a former president of the Microscopy Society of America, a former councilor of the Materials Research Society, a fellow of the American Ceramic Society, and he chairs the British Marshall Scholarship program's North-Eastern Regional Selection Committee. Dr. Hobbs received a B.Sc. degree in materials science from Northwestern University and the D.Phil. degree in science of materials from Oxford University as a Marshall scholar.

Edward Lahoda is an advisory engineer at the Westinghouse Electric Science and Technology Department. He has more than 25 years of experience in process analysis, development, design, and field support. He has extensive background in the manufacture of uranium-based fuels and operation of the waste treatment and other ancillary systems. In the environmental area he was responsible for the technical development and field startup of the Westinghouse soil washing and high-temperature thermal desorption technologies. He has chemical process design experience in processing chemical warfare agents, nuclear fuels, and high-and low-level nuclear wastes and in plasma processing of wastes and plasma production of specialty materials. He has served on committees at the Savannah River Site addressing overall operation and test data validity of the Defense Waste Processing Facility, chaired the In-Tank Precipitation Chemistry Review Panel, and was a member of the In-Tank Precipitation Replacement Review Panel. He served as a technical expert for the National Research Council Committee on Alternative High-Level Waste Treatment at the Idaho National Engineering and Environmental Laboratory. He is a member of the American Institute of Chemical Engineers. Dr. Lahoda received his B.S.,

M.S., and Ph.D. degrees in chemical engineering from the University of Pittsburgh and his M.B.A. from the University of Pittsburgh.

Robin Rogers is a professor of chemistry and director of the Center for Green Manufacturing at the University of Alabama. Dr. Rogers' research interests include green/sustainable separation science and technology, aqueous biphasic systems, room temperature ionic liquids, environmentally benign polymer resins, crystal engineering, and radiochemistry. He is a member of the American Chemical Society, American Nuclear Society, American Crystallographic Association, and Sigma Xi. Dr. Rogers is also the editor of the American Chemical Society journal Crystal Growth and Design. Dr. Rogers received his B.S. and Ph.D. in chemistry from the University of Alabama and reached the rank of presidential research professor at Northern Illinois University.

Ben Sternberg is a professor and director of the Laboratory for Advanced Subsurface Imaging at the University of Arizona. Dr. Sternberg's research interests include high-resolution geophysical surveys, instrumentation, data acquisition, data processing and interpretation, and subsurface imaging including image processing and pattern recognition of geophysical signatures. He has 10 years of industry-related experience prior to his academic appointment. Dr. Sternberg has been elected president of the Near-Surface Geophysics Section of the Society of Exploration Geophysics and was appointed chair of The Technical Academic Review Group for the U.S. Department of Energy to review geophysics projects. Dr. Sternberg also is serving on the National Research Council's Committee on Seeing into the Earth. He received his B.S. in physics and his M.S. and Ph.D. in geophysics from the University of Wisconsin, Madison.

Edwin L. Zebroski consults on decision analysis and risk management in the nuclear and chemical industries and for several national laboratories. Previously he led large-scale development efforts in materials, chemical processing, fuel cycle, and reactor design of commercial and naval nuclear power systems. He has written more than 150 technical publications, including several patents and sections of six books. He has served on panels for the National Research Council, the Department of Energy, the Department of Interior, the National Science Foundation, the former Commission on Engineering Education, and the Committee on Prioritization and Decision Making in DOE-OST. Dr. Zebroski is a member of the National Academy of Engineering, with a B.S. in physics and chemistry from the University of Chicago, and a Ph.D. in physical chemistry from the University of California, Berkeley.

Appendix F

List of Presentations

MARCH 30, 2000
Washington, D.C.

Science and Technology Projects and R&D Portfolio, Mark Gilbertson, DOE-EM

Needs and Opportunities for High-Level Waste Research, Mark Gilbertson, DOE-EM

EMSP Overview, Mark Gilbertson, DOE-EM

High-Level Radioactive Waste Complex Overview, Ken Picha, DOE-EM

Tanks Focus Area: Introduction and Overview, Kurt Gerdes, DOE-EM

JUNE 12, 2000
Richland, Washington

River Protection Project Overview, Don Wodrich, DOE-Office of River Protection

National Academy of Science Review of High-Level Waste Long-Term Needs, Billie Mauss, Tanks Focus Area, DOE-Richland Operations Office

High-Level Waste at the Idaho Site, Joel Case, Idaho National Engineering and Environmental Laboratory, DOE-Idaho Operations Office

Technical Review of Richland Operation Office Waste Treatment Process and Facility, Neil Brown, DOE-Richland Operations Office

AUGUST 28, 2000
Augusta, Georgia

High-Level Waste Overview, Jerome Morin, Westinghouse Savannah River Company (WSRC)

Canyon Processing at Savannah River Site, Frank Graham, Savannah River Technology Center

West Valley Demonstration Project HLW Processing Experience, Steven M. Barnes, Washington West Valley Nuclear Services Company

Long-Term Research Needs to Support Storage and Removal of High-Level Wastes and Pretreatment of High-Level Waste Sludges, David Hobbs, Paul d'Entremont, Brenda Lewis, Eloy Saldivar, Bill Wilmarth, Phillip Zapp, Savannah River Operation Office

Long-Term Research Needs for Salt Processing, Samuel Fink, Joe Carter, David Hobbs, Reid Peterson, Mark Barnes, Doug Walker, Savannah River Operation Office

Vitrification Science Needs, Sharon Marra, Ned Bibler, Richard Edwards, Carol Jantzen, Joe Ortaldo, Savannah River Operation Office

Saltstone and Effluent Treatment Facility Long-Term Research Needs for the National Academy of Sciences, Elmer L. Wilhite, Arthur W. Wiggins, Daniel J. McCabe, John R. Fowler, Tom Lookabill, Savannah River Operation Office

Appendix G

Glossary

Anisotropy: See the definition of Isotropy.

Batch: The word "batch" in glass technology is used conventionally in two ways: *batch materials* are precursor materials that, when heated in the melter, react to form the molten glass; a *batch process* is one that operates on an intermittent or interrupted basis (i.e., is not continuous) and in which the processing vessel (e.g., the melter) is emptied before being replenished.

Becquerel: The international unit of radioactivity, equal to one disintegration or other nuclear transformation per second. The historical unit of radioactivity is the curie, equivalent to 3.7×10^{10} becquerels.

Bismuth phosphate process: This process separated plutonium from uranium and other radionuclides in spent fuel. It operated on an industrial scale at the Hanford Site from 1944 to 1956. Following dissolution of the aluminum cladding in hot sodium hydroxide solution, the uranium metal fuel slugs were dissolved in nitric acid. The dissolved fuel solution was treated with several successive cycles of precipitation to separate plutonium from uranium and other elements. The first involved inducing precipitation with bismuth phosphate by way of co-precipitation with lanthanum fluoride. When sufficiently concentrated, the plutonium was separated from the lanthanum fluoride carrier. The acidic HLW from this process contained uranium, fission products, and a significant fraction of the plutonium. It was made alkaline with sodium hydroxide prior to storage in tank farms.

Calcination: Decomposition of a material to a powder, usually involving an oxidizing atmosphere at an elevated temperature, for the purpose of changing its properties (e.g., to dehydrate or to cause thermal decomposition of nitrates, nitrites, oxalates, formates, and similar anions).

Calcine: General term for the powder generated by calcination. For the INEEL HLW program, it is a specific product from a fluidized bed calciner.

Curie: Measure of radioactivity equal to 3.7×10^{10} disintegrations per second, an historical unit representing approximately the radioactivity of 1 gram of radium-226. The Système International or SI unit for radioactivity is the becquerel, 1 disintegration per second.

Decontamination factor: The (dimensionless) ratio of the concentration of a species of interest in an original, input stream (to a process) to the concentration in a final, output stream. The typical quantity used for this ratio is the radioactivity per unit volume or mass of carrier (e.g., solvent) material.

Electrode: The conducting material used in electric melters to pass an alternating current through the preheated glass batch so as to cause internal resistive heating and melt formation. Electric melters for commercial glass production typically employ molybdenum or tin oxide electrodes. The Joule-heated melters at DWPF and WVDP use Inconel-690® alloy electrodes. Inconel-690® is a registered trademark of Inco Alloys International Inc., Huntington, West Virginia.

Elution: Flushing of a solvent through a sorbent material to eliminate the adsorbed component.

Environmental Assessment (EA) glass: Alkali borosilicate glass developed at the Savannah River Site, used in the Environmental Assessment of the DWPF, and specified in the SRS Waste Acceptance Product Specification (WAPS) as a benchmark for comparison, based on the durability test. The WAPS requires that HLW glasses be more durable (in the Product Consistency Test leach test analyses) than EA glass.

Foaming: Result of redox reactions and the breakdown of anions (nitrates, carbonates, etc.) that generate gas during melting. Excessive foam can form a physical and thermal barrier between the cold cap and the melt pool and can lead ultimately to melter shutdown.

Frit: Premelted glass that has been crushed to a fine powder and used for further ceramic processes (e.g. ceramic glaze formulation). The frits used for HLW vitrification are premelted borosilicate glass powders that are added to the waste and fed to the melter.

Glass-in-glass phase separation: Phenomenon that occurs when a melt separates into two or more amorphous phases during cooling. Depending on the mechanism and the relative volumes of the separated phases, the morphology may vary from encapsulated droplets in a continuous matrix to completely continuous inter-dispersed phase regions. The concern with borosilicate systems is that phase separation could lead to the formation of borate-rich glass regions of poor durability.

Gray: The unit of absorbed dose equal to the absorption of one joule per kilogram of absorbing material.

High-level waste: High-level waste can be (1) irradiated reactor fuel; (2) liquid waste resulting from the operation of the first-cycle solvent extraction system, or equivalent, and the concentrated waste from subsequent extraction cycles, or equivalent, in a facility for reprocessing irradiated reactor fuel; and (3) solids into which such liquid wastes have been converted (Title 10 CFR Part 60.2). In loose terms, HLW is the waste containing fission products and actinides that results from the reprocessing of spent nuclear fuel and requires permanent isolation in a geologic repository.

Ion-exchange: Chemical reaction in which mobile hydrated ions of a solid are exchanged, equivalent for equivalent, for ions of like charge in solution. One or more of the ionic species in solution are selectively absorbed on a solid and retained.

Isotropy: The feature of having equal physical properties (refractive index, thermal expansivity, elastic constants, etc.) in different crystallographic directions. Cubic crystals are rarely completely isotropic, but are considerably more so than other crystal systems (tetragonal, hexagonal, orthorhombic, monoclinic or triclinic) with lower symmetry, which are usually strongly anisotropic.

Leaching: Chemical dissolution of a solid in a liquid, which may be congruent (i.e., the solid dissolves uniformly) or incongruent (chemical components of the solid are dissolved at different rates). Borosilicate waste glasses are required to be leach resistant with respect to aqueous liquids (deionized water or groundwater) and usually exhibit incongruent leaching, with the most mobile ions being the alkali ions (including cesium) and boron.

Low-level waste: Any radioactive waste that is not spent fuel or high-level waste.

Methathesis: Double decomposition chemical reaction of the type $AB + CD \rightarrow AD + CB$ that is driven by the law of mass action when there is an excess of one ion.

Mixed waste (MW)**:** Waste that contains both chemically hazardous constituents regulated under RCRA and radioactive materials regulated under the Atomic Energy Act.

Off-gas emissions: Gaseous and particulate material that is vaporized from the surface of the melt pool in a glass melter.

Partitioning: Process that divides an input stream into two or more output streams. A mixture of solids can be partitioned based on differences in material properties. A liquid phase with two or more chemical species can be partitioned using selective media to segregate the species based on their differences in chemical reactivity.

Phenomenological model: A multi-component model with predictive power that combines a series of mathematical descriptions of the individual phenomena involved. Some of these processes occur simultaneously, others occur sequentially. In the case of wasteform leaching, these will probably include some or all of the following:

- H^+ or H_3O^+ diffusion into the glass surface;
- diffusion of leachable species (particularly alkali ions) out of the glass surface;
- subsequent diffusion of ions across the external leached layer;
- dissolution of the leached layer itself (i.e., matrix dissolution);
- processes (e.g., fracturing) that alter the glass surface-area/solution-volume ratio (SA/V ratio);
- precipitation reactions on the glass surface from interaction with the chemical components of groundwater;
- groundwater radiolysis and its effect on wasteform/groundwater interactions;
- internal radiation effects on the structure and durability of the wasteform.

Product Consistency Test (PCT): The product consistency test is a leach test that evaluates the chemical durability of glasses by measuring the concentrations of the chemical species (boron, sodium, and lithium) released from a crushed glass to a test solution. The PCT is one of the standard tests of the American Society for Testing and Materials (C1285-97).

PUREX Process: This process purified and separated both uranium and plutonium from spent fuel. It was used at all four HLW sites and first used at Hanford in 1952. The first step consists of removing the cladding of spent fuel (fuel preparation) to expose the fuel rods. In the second step, spent fuel is dissolved in nitric acid. The third and key step is the separation and recovery of uranium and plutonium from other fission products, which is done in a continuous counter/current solvent extraction process using tri-N-butyl phosphate in a hydrocarbon diluent (such as kerosene). Plutonium and uranium remain dissolved in the aqueous nitric solution as nitrates. This step produced the majority of HLW, consisting of a nitric acid solution of fission products and impurities. The fourth step consists of separating plutonium from uranium followed by conversion to solid oxide or metal. Except at INEEL, the waste was neutralized with sodium hydroxide prior to storage in the tank farms. A major advantage of the PUREX process over previous processes used (see Bismuth phosphate process and REDOX process) is that nitric acid is the only salting agent used in the aqueous phase; therefore HLW has a lower salt content than waste from the other processes. Other advantages of PUREX are reduction of the waste volume, greater flexibility in process conditions, reduction of fire and other hazards, and reduction of costs. Processes based on PUREX have been adopted for nearly all fuel reprocessing throughout the world.

Rad: Radiation absorbed dose; the basic unit of absorbed dose equal to the absorption of 0.01 joules per kilogram of absorbing material.

Radiolysis: The chemical modification or decomposition induced by ionizing or near-ionizing radiation. In the case of borosilicate glasses, radiolysis can lead also to atomic displacement.

Raffinate: Aqueous solution remaining after a metal has been extracted by the solvent.

REDOX process: The REDOX process purified and separated both uranium and plutonium from spent fuel. It was operated at Hanford from 1951 to 1967. This was a continuous counter/current solvent extraction process using methyl isobutyl ketone (hexone) as the solvent. With hexone, it was necessary to add aluminum nitrate to the aqueous phase to achieve the high ionic strength required for effective extraction of the uranium and plutonium. This process had major advantages over the bismuth phosphate process: it reduced the waste volume, provided for nearly complete recovery of both uranium and plutonium, and allowed continuous operation, which favored higher production rates. One disadvantage was that large amounts of aluminum nitrate had to be added. The acidic waste, containing fission products, aluminum nitrate, and impurities, but very little uranium or plutonium, was made alkaline with sodium hydroxide prior to storage in the tank farms.

Refractories for glass melters: The high-temperature materials used for thermal insulation and for containing molten glass. The glass contact refractories are selected for high corrosion resistance, while the backing refractories are chosen for their insulating properties. Refractories used in electric melters must also exhibit high electrical resistivities. The glass contact refractories used in the Joule-heated melters at DWPF and WVDP are bricks of chrome-rich Monofrax K3®, a fusion-cast refractory material that shows excellent corrosion resistance, but with a trend toward decreasing electrical resistivity with increasing temperature that can be problematic much above 1150°C. Monofrax is a registered trademark of the Carborundum Company, Falconer, New York.

Repository: A place for the disposal of immobilized high-level waste to isolate it from the environment.

Reprocessing: Recovery of fissile and fertile material for further use from spent fuel by chemical separation of uranium and plutonium from other transuranic elements and fission products. Selected fission products may also be recovered. This operation results in the generation of HLW.

Saltcake: concentrated waste in the form of crystallized salts resulting from the evaporation of liquid high-level waste.

Saltstone: Low-level wasteform material used at the Savannah River Site, consisting of a cement-based material in which the low-level waste is either dissolved or dispersed.

Sludge: Semi-solid, viscous, amorphous material that is substantially insoluble. The sludge in HLW tanks is formed by gravity settling of insoluble solids (usually hydroxides) on the tank floor after waste neutralization with sodium hydroxide.

Slurry: Suspension of insoluble (usually finely powdered) materials in a liquid medium.

Solvent extraction: Separation of materials of different chemical types and solubilities by selective solvent action. This term is a general process for separating one or more chemicals solubilized in one solvent by the use of a second solvent that is (1) insoluble in the first solvent and (2) selective in its ability to bind one or more of the chemical constituents. After co-mixing and agitation of the two solvents, they separate, with the second having extracted the chemical species of interest.

Source term: The initial amount of radioactivity used to calculate exposure and doses to various receptor groups.

Spent nuclear fuel: Fuel that has been withdrawn from a nuclear reactor following irradiation, the constituent elements of which have not been separated by reprocessing.

Spinel: Family of cubic-structured ceramic oxides of general formula $A^{2+} B_2^{3+}O_4$. The spinel compounds that can form precipitates in borosilicate glass during high-level waste vitrification are predominantly oxide compounds of iron, nickel, and chromium.

Transuranic waste: Waste contaminated with transuranic elements with half-lives greater than 20 years, in concentrations greater than 100 nanocuries per gram.

Waste loading: Weight fraction of waste immobilized in the waste form (e.g., borosilicate glass). Some ambiguity has been caused by the use of the term to describe either the weight fraction of the original waste components (salts, hydroxides, etc.) in the waste form or the amount of equivalent oxides that will be formed by thermal decomposition during waste form production. A further ambiguity arises from use of the term at different HLW sites to either include or exclude nonradioactive waste components (e.g., Na, Al, Si) that constitute part of the final waste form matrix.

Appendix H

List of Acronyms

CFR	Code of Federal Regulations
D&D	Deactivation and Decommissioning
DF	decontamination factor
DOE	U.S. Department of Energy
DST	double-shell tanks
DWPF	Defense Waste Processing Facility
EA	Environmental Assessment
EM	DOE Office of Environmental Management
EMSP	DOE Environmental Management Science Program
EPA	U.S. Environmental Protection Agency
ESPIP	Efficient Separations Program-Integrated Program
HLW	high-level waste
INEEL	Idaho National Engineering and Environmental Laboratory
LLW	low-level waste
LWR	light water reactor
MW	mixed waste
NAE	National Academy of Engineering
NAS	National Academy of Sciences
NASA	National Aeronautics and Space Administration
NM	nuclear materials
NRC	National Research Council
OCRWM	DOE Office of Civilian Radioactive Waste Management
OST	DOE Office of Science and Technology
PAR	Process Acceptable Region
PCT	product consistency test
PCCS	Product Control Composition System
PGNA	Prompt Gamma Neutron Activation
PPM	part-per-million
PUREX	Plutonium and Uranium Recovery by Extraction

RCRA	Resource Conservation and Recovery Act
REDOX	reduction and oxidation
RFP	Request for Proposals
SBW	sodium-bearing waste
SC	DOE Office of Science
SITE	Superfund Innovative Technology Evaluation
SLS	solids-liquid separation
SNF	spent nuclear fuel
SREX	strontium extraction
SRS	Savannah River Site
SST	single-shell tanks
STCG	Site Technical Coordinating Group
TBP	tributyl phosphate
TFA	Tanks Focus Area
TOC	Total Organic Carbon
TPA	Tri-Party Agreement
TPB	tetraphenylborate
TRU	transuranic
UNEX	Universal Solvent Extraction
USNRC	U.S. Nuclear Regulatory Commission
WAC	Waste Acceptance Criteria
WAPS	Waste Acceptance Product Specification
WIR	waste incidental to reprocessing
WSRC	Westinghouse Savannah River Company
WVDP	West Valley Demonstration Project